Product data exchange

Product data exchange

M. Susan Bloor and Jon Owen

University of Leeds

UCL

PRESS

First published in 1995 by UCL Press

UCL Press Limited
University College London
Gower Street
London WC1E 6BT

and
1900 Frost Road, Suite 101
Bristol
Pennsylvania 19007-1598

British Library Cataloguing in Publication Data
A catalogue record for this book is available from the British Library.

Library of Congress Cataloging-in-Publication Data

Bloor, Susan, 1941–
Product data exchange / Susan Bloor and Jon Owen.
p. cm.
Includes bibliographical references and index.
ISBN 1-85728-279-5 (hb)
1. Computer-aided design—Standards. 2. New products—Databases.
3. Computer-aided engineering. I. Owen, Jon, 1957– . II. Title.
TA174.B57 1994
620'.0042'0285—dc20 94-42780
CIP

Typeset in Zapf Elliptical.
Printed and bound by
Biddles Ltd., Guildford and King's Lynn, England.

Contents

Preface vii

The authors xiii

Part I
First-generation methods for CAD data exchange: the *ad hoc* solution 1

1 Introduction 3

2 CAD and CAD data exchange 9

Part II
Second-generation methods: the neutral format 25

3 Early versions of IGES (1979–1983) 27

4 The proliferation of product data exchange initiatives 51

5 Further developments 63

6 Related initiatives 71

Part III
Complementary initiatives and infrastructure 83

7 Problems with product data exchange 85

8 Testing 105

9 Subsets and application protocols 121

10 Conformance testing 133

Part IV
The third generation: STEP 151

11 The history of STEP and associated activities 153

12 A brief description of STEP 169

13 The way forward 191

Appendices

A Abbreviations 205

B Standards bodies 209

C Enabling technologies 211

D IGES coverage 215

E IGES entity prescriptions 223

F VDA-FS entity prescriptions 237

G SET sub-block prescriptions 239

H Addresses and telephone numbers 245

Textual notes 247

References 249

Index 259

Preface

This book charts the developments in product data exchange from its inception to the present time. There has been a need to exchange data which describes a product – often its geometric shape – since computers were first used for design and analysis in the 1950s. That need still exists today, with the data needing to be available throughout the life-cycle of the product and throughout the supplier chain.

The treatment is chronological, based on three generations of methods for effecting product data exchange. It concentrates on developments since the publication of IGES in 1980 (see Fig 0.1). Although there are still many problems to be solved, the publication of STEP as an international standard in 1994 seems to mark a better time than most to try and document the tremendous progress and increased understanding that has been achieved over the past few decades.

The book begins by charting the development of computer aided design since the 1950s, because the design phase is where the product definition evolves. The early use of computers for analysis and visualization and representation (of sculptured surfaces) contributed heavily to the "islands of automation". Chapter 2 describes the product life cycle and the nature of design which led us to require easy information exchange and thus makes the case for exchanging product data in computer-readable form. It reflects the understanding up until the late 1980s that shape-definition data was overwhelmingly important.

There are several approaches to the problem ranging from imposing the same CAx (computer-aided system) on all members of one's consortium, through graphical representation of drawings and *ad hoc* software. The one we wish to espouse, except in special circumstances which we describe, is that of neutral formats. The *ad hoc* route can be considered, as it is by the authors, as a first-generation approach. The second half of

	1980	1981	1982	1983	1984	1985	1986	1987	1988	1989	1990	1991	1992	1993	1994
STEP					STEP→				DPI	initial parts				11, 31 approved	initial release
US					PDDI		PDES Initiation		PDES Inc →						
IGES	IGES 1.0			IGES 2.0			IGES 3.0		IGES 4.0		IGES 5.0	IGES 5.1		IGES 5.2	
Subsets								MIL-D-28000					MIL-D-28000A		
Germany					VDA-FS 1.0		DIN	VDA-IS 1.0 VDA-FS 2.0 VDA-PS		VDA-IS 2.0					
France					SET 1.1	afnor				afnor					
Europe					ESPRIT→					ESPRIT II→				ESPRIT III→	
Testing		Autofact→				CTS→	NAVFAC			CTS-2→					
NEDO					NEDO1		NEDO2								
Graphics						GKS		CGM	GKS-3D	PHIGS		CGI	PHIGS-PLUS CGRM		
EDIF						EDIF 1 0 0		EDIF 2 0 0						EDIF 3 0 0	
Modelling		IDEF0-2	NIAM			IDEF1X							IDEF3-4		

Figure 0.1 Chronology of developments in product data. An arrow indicates the start of a programme of research or a particular project. Entries which do not have an arrow are placed in the column for the year in which the specification or standard was published: work would necessarily have been undertaken before that date, but is not shown on the chart in order to reduce its complexity.

Chapter 2 shows why the route using neutral formats is the one that deserves to be followed. The chapter makes brief mention of vendor-specific neutral formats such as ISIF (Intergraph), IIF (IBM) and the ubiquitous DXF (Autocad). The chapter ends with a description of the standards-making process, since this has a large impact on the time to produce a standard but should guarantee that what is produced is worthwhile.

We designate as "second-generation specifications" those formats which predate the ISO initiative started in 1984 to produce an internationally accepted standard. We would also define them as CAD data exchange formats and their content is largely geometrical. Part II is dedicated to IGES, responses to IGES from both the users and the specification community, and also other specifications relevant to communications in computer-aided engineering. Chapter 3 deals with IGES developments between 1979 and 1981 including its history, the development process and the attempts to demonstrate the technology. The difficulties, both real and perceived, are addressed. These led to alternative specifications, described in Chapter 4, which also led to user actions. These included the establishment of guidelines and understanding of the problem, which is that data exchange software processors are not "black-box" or "appliance" technology. Chapter 4 covers VDA-FS 1.0, SET, XBF and PDDI. We emphasize the requirements which drove the development of each, the application area addressed by each, and particularly the lessons learned and put into practice in each specification.

Chapter 5 follows the development of IGES, SET and VDA-FS to the present. Chapter 6 covers many related specifications briefly, so that their relevance to product data exchange is established. This includes work in the areas of OSI (MAP, TOP), electrical and electronic engineering (EDIF, VHDL), office documents (ODA, SGML), invoicing (ODETTE), and graphics (GKS, PHIGS, GKS3D, CGM). Graphics standards are important because it is essential to characterize the difference between exchanging product data and one particular manifestation of it, an engineering drawing. Chapter 6 identifies clearly the difference between a graphical representation and a product model.

Part III addresses complementary initiatives and starts in Chapter 7 by identifying fully all the problems with data exchange. It characterizes those problems which are common to any form of product data exchange (such as trying to transfer information between CAD systems which cannot both handle the same information) and those which arise from use of a particular mechanism. Solutions and advice are given for the latter class. The former category cannot yet be solved elegantly. Chapter 8 introduces the very large subject of testing. Many users have instigated testing of all IGES files they receive. The Autofact demonstrations wished to show that IGES could work satisfactorily. Both these activities need software tools, particularly ones which check the syntax and semantics of IGES files. The sorts of testing and the tools produced are described in order to demonstrate the large amount of interest and work (and, unfortunately, duplication of work) that occurred.

It was soon perceived that exchange could only occur if constraints were imposed on use of a specification. The notion of subsets of entities from a specification was taken from GKS. These subsets would support a particular application. Chapter 9 covers the development of MIL-D-28000 in the United States, and work in the IGES/PDES arena which is developing application protocols. These not only give a list of IGES entities but also provide an information model to show how they should be used to capture the information of interest. The subsets of IGES (VDA-IS) defined for use by the German automotive industry and particularly the specific requirements that are placed on processors by VDA-IS specification are presented. The prescriptions for entity conversion imposed by VDA-IS are an important part of the specification.

The foregoing initiatives set the scene for formal testing procedures. The first of these is conformance testing. Chapter 10 categorizes levels of testing as currently understood in order to better explain conformance testing. The chapter covers the scope, how conformance testing can be performed, the problems with current exchange specifications such as ambiguity, and the software tools needed to do conformance testing. The overall methodology which owes much to the lead being given by those conformance testing OSI standards and depends heavily on the concept of abstract test cases, is fully described and justified.

And so to STEP. Part IV, which addresses the third generation of activities, deals with STEP in its entirety. It starts in the mid-1980s, when the first PDES document was published, ESPRIT project CAD*I began and the ISO work to develop an international standard for product data was gaining momentum. We intend that this should give insight into the intent behind STEP, its development story, the technologies required, and what the initial release looks like. Chapter 11 contains the history of its development. Chapter 12 describes the structure of the standard and gives a brief overview of its contents at the time of writing. It does not attempt to do this in any depth for three reasons. First, such a description would soon become out of date, because STEP is now expanding rapidly. Secondly, because of the scope of STEP, the material would fill several books rather than a part of a single one. Finally, the development method is at least as important as the standard itself, and will provide benefits to industries which adopt such methods in their own use of product data: this is addressed in Appendix C.

The final chapter, Chapter 13, concerns the future. Plans for future developments in STEP are presented: application protocols in the pipeline and plans for conformance testing and processor support tools are all covered. The wider issues of STEP's influence, already, on engineering data management methods and on CAx systems are also discussed in the concluding chapter.

This subject like many others is riddled with acronyms. We hope their use does not upset the reader unduly and we have tried to make a comprehensive list in Appendix A. We have also tried to explain terminology as it occurs, but have provided an index which indicates the page on which a term is explained, even if it is not given a formal definition.

In the References, except where cited from within the text, references for STEP have not been included explicitly; the latest documents are given in Table 12.1.

Except where stated explicitly, the status of committees and their leaders, and documents – particularly STEP parts – are those at the time of writing. These will necessarily change, but are included to give at least a snapshot.

The emphasis on IGES and current specifications may be criticized. The authors do believe that these will still be in use in the next century, and this is a view shared by others [165]. STEP is an exciting development which is already contributing to the understanding of engineering data which must be managed if information systems are going to contribute fully to competitiveness. No standard is ready to use until good processors and accredited testing services are in place. These only became available for IGES, SET and VDA-IS relatively recently. The necessity for these and their common use are not yet widely understood, especially by vendors. Although the rationale behind STEP is to facilitate its introduction, STEP is not yet agreed and has gaps we do not yet know how to fill.

There is also a bias towards product data exchange for mechanical engineering applications, rather than for civil, chemical, electrical or electronic engineering. This reflects not only the environment in which we work, but that the use of product data is well advanced in this particular area. However, lessons learned from these other areas – particularly the electrical and electronic – are described.

The book provides a historical perspective on developments in product data exchange. It will be useful to those who are defining the current generation of such standards, and to those who prescribe how any product data exchange specification is used in an industrial enterprise. By describing the various problems which can occur with product data exchange, and some of their solutions, it will also help those involved in the day-to-day use of product data exchange.

We hope this book will excite you and encourage you to think and perhaps volunteer to help with the definition of standards and methods. Even if it does not do this, at least it may encourage patience and deeper consideration before you introduce computer solutions in your own enterprise.

Acknowledgements

The writing of this textbook would not have been possible without the help, support and expertise of many colleagues. We should like to thank the following: those involved in the development of STEP; partners in various national and European R&D projects; members of the British national standards committee BSI AMT/4; the Computer Aided Engineering Group in the Department of Mechanical Engineering at the University of Leeds; colleagues (both present and past) at CADDETC; Dr Hilary Kahn, from the University of Manchester, for her contribution on EDIF; the Design Council for permission to reproduce Figure 2.1; the permission of ISO to reproduce some of the figures from the STEP parts; the reviewers of the text, who made many helpful suggestions, cited references of which we were unaware, and whose constructive comments were much appreciated.

The text will necessarily have omissions, which are neither deliberate nor wilful, for which we apologise. If you have information on a particular project or initiative which has been omitted, then please contact the authors at the following address:

Dr Susan Bloor and Dr Jon Owen
Department of Mechanical Engineering
The University of Leeds
Woodhouse Lane
Leeds
LS2 9JT
United Kingdom

Phone: + 44 113 233 2113
Fax: + 44 113 233 2150
Email: `ISO@.leva.leeds.ac.uk`

The authors

Susan Bloor gained her degree in Mathematics in 1962 and her PhD in 1965. She is Senior Computational Advisor in the Department of Mechanical Engineering at the University of Leeds. Her research interest is in engineering product data modelling and its role in systems integration. She has played a leading role in the department's research activities in this field for many years through a continued involvement in the industrially sponsored Geometric Modelling Project, as Principal Investigator on Teaching Company Schemes, and as Project Manager for various European-funded projects.

After gaining both a BSc and a PhD at the University of Leeds, *Jon Owen* spent two years as a lecturer in the Department of Computer Studies and five years in the Geometric Modelling Project in the Department of Mechanical Engineering. In 1986, he joined CADDETC and for six years led and co-ordinated the research and development activities. He is now a Senior Fellow in the Department of Mechanical Engineering. Since 1985, he has been active in STEP development and a member of BSI-AMT/4. He was one of the original seven members invited to participate on the European Commission's Product Data Technology Advisory Group (PDTAG).

Part I

First-generation methods for CAD data exchange: the *ad hoc* solution

Part I explains the engineering context for the exchange of computer-aided design data. In particular, it explains how the need has arisen and the early *ad hoc* attempts to solve the problem. The terminology that has grown up around CAD data exchange and product data technology is also introduced.

Chapter 1

Introduction

1.1 The problem

The problem of product data exchange addressed in this book is due to the way in which the use of computers has been introduced in industry. Manufacturing companies from all sectors have been using computers since the early 1950s to do operations where their use was advantageous in some way. Over that time, there has been increasing technological complexity in both products (e.g. aeroplanes) and manufacturing methods (e.g. the mass production of well designed consumer goods). The requirements for analysis of design, for accuracy of manufacture and for competitive products were all increasing. In parallel, human demands were changing. People were not as willing to do repetitive jobs, organizations were getting larger and the customer/supplier chains were growing. Even without hindsight, one can see the attraction of computers. For example, they could do a stress analysis for the aircraft industry, which is a tedious, long and hence error-prone job by hand. By the late 1950s, generation of numerical control codes for machine tools and visualization for designers were becoming feasible. Data storage and processing of records helped administrative and business functions, payroll, stock control, sales and invoicing.

Computers and information technology have been introduced into industry in an *ad hoc* manner to relieve particular bottlenecks in individual processes. There was no need to think of the effect on the overall enterprise – today, we would suggest that an information technology strategy should support a business strategy, and hence integration is built in. However, since the introduction of computers, there has been an increased understanding that different processes use the same information (e.g. geometry). Consequently, computers were storing duplicate information in different forms. Additionally, to facilitate the use of such data, its transfer became an issue. Integration of computer applications became important.

The objectives of both integration and transfer have been elusive. Product data exchange is the enabling technology that has evolved to support these issues, particularly for engineering data about products. The essence of the problem is how to translate between different data forms and that is the subject of this book.

"Product data" is a term which has changed its meaning over the last ten years to describe *all* data relating to a product. The book addresses product data exchange: product data is all the data that represents a product. The term "product model" implies a simulation of the product as in a physical or mathematical model. The intention is to interpret the product data to reproduce important features in the product. If this can be done, we have a computer model. Product model data is synonymous with product data in STEP, the proposed standard for the exchange of product model data.

The aforementioned are instances or representations of a product in the computer. The term "product data model" is analogous to a schema and describes the *form* of the data used to represent a product. "Product definition data" is that captured explicitly or implicitly in an engineering drawing.

1.2 Business and engineering

There are two strands that can be identified. One is the progressive computerization of the business functions and the other is that of the engineering function. By this latter we mean the development and manufacture of a product from concept to maintenance and, if relevant, decommissioning. Business functions concentrated first on database management systems and, later, on management information systems. Engineering used number-crunching, graphics and generation of control data (e.g. for machine tools). This book concentrates on engineering, but not primarily on the applications and technologies just mentioned. The way in which automation technology has been applied has left us with a legacy of consequent communications problems, and that is the emphasis we choose [42].

1.3 Communication

It is a truism that all organizations and groups of organizations depend on effective communication: design to manufacture, specifications to suppliers, specifications to customers, time-keeping to payroll, delivery dates to sales; the examples are legion because communication *is* the organization. Unfortunately, the installation of computer aids, in an *ad hoc* manner as mentioned above, has arguably worsened communication. Admittedly, the scale of operations is larger than in pre-computer days, and the age-old methods used by humans (talking, writing and drawing) would not now cope; but they have not been replaced. A computer-aided draughting system can be used to produce a detail drawing, which will mostly be for the production section. Then data will have to be typed from that drawing into the numerical control program after choosing that part of the information on the drawing that is relevant and reformatting it as necessary. This sort of operation is performed throughout, outputting data from one program and inputting in another form to other programs. Thus new time delays and new sources of error have been created.

Computer communications have many facets and, to aid understanding, the analogy with telephone communication is often used. The requirements for making a successful contact are that one's message is understood fully by the person to whom one is talking, and that it is the correct person. There is no need to check or repeat it in writing if the information has been conveyed already. The possibilities for communicating the message include telephones and connection to networks (for both persons), which means knowing each others' numbers, the availability of lines and the name of the person to whom you wish to talk. You also need to speak a common (natural) language, possibly including technical terms, and have common experience. These can be categorized as the medium (how), the message (what), and the context (for what purpose). Although rarely optimal in terms of time, we can be satisfied in many cases that it is possible to make the connection using magnetic tape or floppy disk. This is achieved by adhering to agreed procedures, as is addressed further in Chapter 7. Others are dealing with telecommunications and networking which will eventually be the common way and save time. Currently, these are too expensive or slow for most engineering communications. Our problem is the message and its context which are essential for effective engineering to support competitive industrial operations.

1.4 Information and data

Data can be transferred when it is a computer representation that the recipient would be able to interpret. The recipient may be a person or a program written by a person with experience and knowledge of the data format. It is much more difficult to transfer the information itself.

"Data" are the details without the context. Adding detail does not turn data into information. Much of the input and output to computer programs is lists or tables of decimal numbers (because people cannot cope with binary). Examples include the values of points and perhaps tangent vectors that lie on a surface, to which people add the context. "Information" is data with a context, or with human interpretation. The words are often used interchangeably, which creates confusion [166].

The problem we wish to solve is transfer of information to support engineering where different divisions, functions and partners have computerized to suit their immediate problem. Engineering drawings are data from which information can be extracted because their context is known and engineers are trained to interpret them.

It should perhaps be stated here that the business functions, particularly order processing and invoicing, have hijacked the term electronic data interchange (EDI). Networks are in place to transfer this information between subscribing parties (which are most large commercial companies) [14].

1.5 Integration and exchange

We have stated that because of the legacy of *ad hoc* application of computers the necessary transfer of information between parties has become more difficult. These parties may be internal to a company (as in the transfer from a draughtsman to a stress analyst) or external (such as supplier chains).

There are two ways of addressing the problem. First, each pair of parties may be considered by studying the information they pass between each other and solving their problem in isolation. Alternatively, all the communicating parties may be addressed and an attempt made to solve the total problem. The first is the problem addressed by product data exchange technology, although of course it proposes general solutions. The second starts to address "integration", an overworked term captured in the phrase "computer-integrated manufacture" that is every company's aim. Integration is more difficult and has two manifestations:

(a) in trying to bring order to information generated by existing disconnected software (programs, applications) and manual methods, which is addressed under the name "engineering data management" (EDM);

(b) what one would do if starting afresh (integration).

This book traces the emergence of data exchange technology. The subject is particularly exciting because the methods being evolved are increasingly recognized as being the core to EDM and integration as well as the next generation of data exchange. This may not be surprising but it is worth remarking that it is the data exchange community which is trail-blazing.

1.6 Life-cycle and the supplier chain

Information has to flow from phase to phase in the life-cycle of a product (from concept, through production, to maintenance or even decommissioning). Even though iterations and feedback from manufacture and concurrent engineering practice do occur, information should develop and increase as the product goes from concept to scrapheap. The information will be selected (the machinist does not need to know where a part fits in an assembly), converted (the product geometry will not be defined as required by the machine tool), and added to (we would like to know actual measurements of the product to test if it is adequate). Alternatively, product information may flow from the designers in one company to those in another, perhaps to add to the design or because the design output of the first (a drawing) is interpreted in the design system of the second. This transfer of information through the life-cycle is slightly different from the transfer within one phase. Exchange technology has found enough problems with the latter, that is, when both parties are interested in the same information, to occupy it fully. Also, because external interactions with contractual overtones are involved, a solution is important. Wilson stated in 1987 that part geometry was the common thread from concept to inspection and assembly and that other information was derived [166]. On that premise, a large part of the life-cycle information could be communicated if complete geometry were satisfactorily exchanged.

1.7 Product data

The current understanding is that although geometry is important, it does not have a special position as far as product representation is concerned. "Product data" is a relatively new term. It contains potentially all the information about a product throughout its life-cycle. It becomes data once there is a formal representation. In engineering such formality was necessary before the advent of computers. Tables of numbers, stylized reports and, most important, engineering drawings are all data. Engineering drawings which embody product definition data are still a major source of information. They are produced and interpreted according to draughting standards to minimize misinterpretation. It is interesting to note that in 1990 the first requirement was still to exchange engineering drawings: the committee which determined the priorities for STEP ratified this as the top priority. Drawings have not yet been replaced by definitions containing more information implicitly (e.g. definitions based on solid models). To use computers as we would wish, much more information is needed in the computer to represent the product. The Product Automation Project's thesis [123] was that automation was not progressing because we did not know how to represent geometry completely. This would enable all sensible questions about an object's geometry to be answered by processing the data one had. The requirements for product configuration data, the structure of an assembly, for material data and for specification data are all examples of what have to be included in any complete product model.

1.8 Conclusion

By looking at the history of computerization, we have attempted to set the scene and explain the importance of product data exchange technology. This is an enabling technology for:

(a) integration;
(b) engineering data management;
(c) product data exchange.

Additionally, as a tool for concurrent engineering, fast access to correct information is a key: product data technology is again at the core.

Chapter 2

CAD and CAD data exchange

2.1 Computer-aided design

We start with design because the evidence is that the information generated during design – the definition of the product – is what people most often wish to communicate. Before computers, the engineering drawing was in universal use. From the early 1950s, it had been stylized and standardized to convey as accurately as possible as much information about the definition of a mechanical product as necessary to suppliers, to customers and to manufacturers. It is a manifestation of product definition data and is produced during the design process. Even so, it should be noted that engineering drawing conventions vary from one country to another.

Other sectors have similar pivotal drawings. The flowsheet, piping and instrumentation diagram, and isometric drawing are examples from the process industry. Electrical schematics are widely used to represent the connectivity in electrical circuits.

From our perspective, that of automation technology, these drawings have shortcomings largely due to the fact that they were developed for production and use by humans. The information about solid objects in an engineering drawing has to be interpreted, as draughtsmen represent them by orthogonal projections: that is, as something that can be drawn accurately. The national draughting standards do not cover sculptured surfaces; they are difficult to draw. However, engineering drawings are still in use and, as has been indicated in Chapter 1, the primary requirement from STEP, as late as 1990, was that it should support transfer of two-dimensional draughting with explicit dimensioning: namely, engineering drawings.

Similarly, other drawings are stylized and can be interpreted fully only by trained people. For example, computer systems are not yet able to represent all of the design intelligence shown in schematics.

CAD is an acronym that is variously interpreted as computer-aided design or computer-aided draughting. As draughting may be considered an aid to design, perhaps it does not matter how it is interpreted. However, in this book, CAD is taken to mean computer-aided draughting which is the use of the majority of CAD systems. CADCAM has been used by the authors, and others, to imply the intent that CAD should facilitate the link with computer-aided manufacturing.

Other things happen during design and Figure 2.1 shows the stages and processes that are considered to make up design [111]. As far as we are concerned the activities of analysis in its many forms (stress, thermal, kinematic, flow), the representation of surfaces, and the production of drawings are emphasized because they were early uses of computer technology. This was because practitioners came up against a need to improve their methods and the use of computers presented the obvious solution.

2.2 Examples of islands of automation

We examine three examples in a chronological perspective to show how computerization has occurred. These are applications which were ready for computerization as the hardware and software became available.

2.2.1 Stress analysis

Stress analysis facilitates the study of how a structure behaves when it is subject to loads and variations in its surroundings. This analysis is necessary to ensure the artefact will be strong enough. Tables and formulae for simple structures such as beams which take into account material behaviour (assuming this is straightforward) have been in existence for a long time. One could do the analysis and add some strength for good measure; these methods are still in use. However, for sculptured shapes, new materials, and machines like aeroplanes and cars where safe operation over a long period with continuously varying loads are the norm we require more complex analyses as well as rigorous testing. Testing is expensive, can be impossible (as for nuclear reactors) and is certainly not practical within an iterative design loop. Analyses often require the iterative numerical solution of partial differential equations. This is time-consuming and tedious and therefore error-prone when done by hand. Assessment of alternatives was no more feasible than by building models. Such number-crunching was the first use to which computers

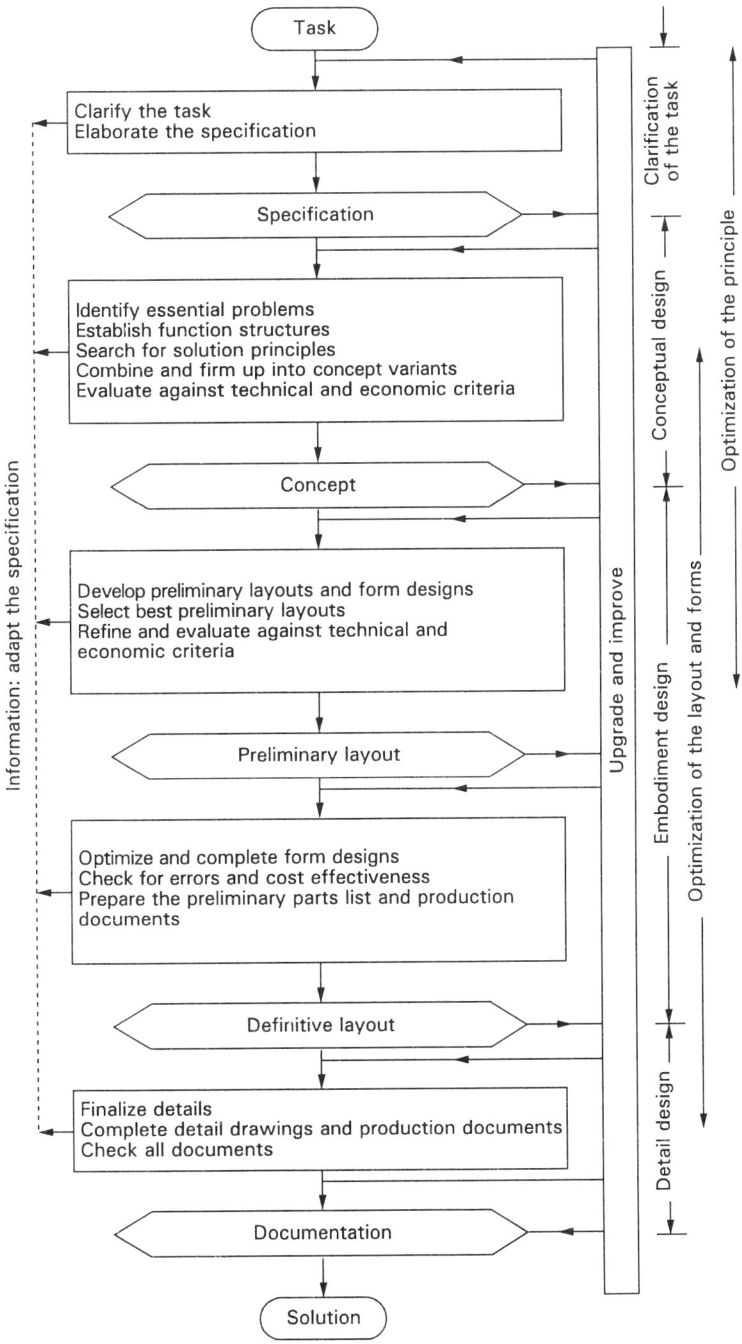

Figure 2.1 The stages of a design [111].

were put. The development of the finite-element technique, which is the basis of innumerable packages for stress analysis, occurred alongside the use of computers for this design function.

These stress-analysis packages need information concerning a representation of the shape of the object to be stressed, any constraints on its movement, and material properties. The method of finite elements also needs the decomposition into elements. In most cases this data is input by hand. Techniques are available for semi-automatic and perhaps automatic meshing, but these require a representation of the geometry of the object. Today this geometry is still the starting point in most analyses even though the geometry will probably be in the computer already, usually in a different form which depends on how the shape has been defined. The results are displacements and stresses. Again these are interpreted manually to check how they might influence the design of the product. Attachment of the results to the master geometry (stored by the computer-aided design system) is not straightforward.

2.2.2 Surface representation

A second application for computers was the representation of complex surfaces. Sculptors could model them and artists could represent them. Skilled people could make moulds by hand so they could be cast. But the concurrent demands of the many for attractive-looking consumer goods, of high technology for lighter, cheaper, faster aircraft (and cars) and the invention of numerically controlled milling machines capable of making accurate sculptured shapes led to a need to represent surfaces numerically. This problem is not yet completely solved since surfaces cannot be truly designed, actual models are still made for visualization at concept design stage and copy-milling has probably only just disappeared. Surfaces are often represented by polynomial interpolation in a network of patches which cover the surface and conform to a given boundary. The amount of data necessary to define a surface and the number of calculations then required to represent it sufficiently finely so that it can be machined accurately are obvious candidates for computer application. These surfaces can then be used to generate control instructions for a machine tool. It also became possible to view these surfaces which were almost impossible to draw adequately for engineering purposes. The hardware to present realistic hidden-surface pictures of rotating sculptured objects is now becoming commonly available. Usually surface packages are self-contained and have been designed to facilitate manufacture of the surface. They are often disconnected from any other part of the product definition (design) process. Several companies quote tens of different packages they need to take a product from concept to realization. Surface data is output frequently only to be re-input in another program in a changed form for some reason.

2.2.3 Engineering drawings

Lastly we come to engineering drawings. They are increasingly complex because less can be left to human experience when products themselves are more complex. There is a long-standing shortage of skilled draughtsmen. Graphics devices became available in the late 1950s and the Tektronix storage tube which drew vectors was ubiquitous in the early 1970s. So, drawing-board analogues began to appear. These allowed the draughtsman to replace his pen by some input device (a combination of text, mouse, thumbwheel or joystick), his drawing board by a screen, and his filing drawer by magnetic file store. Many computer-aided draughting programs were marketed and the user interfaces to these have continued to be improved. With one hard-copy device in the drawing office, engineering drawings could be produced more quickly and neatly. But there was no additional intelligence in the result. Surface systems rarely contributed to engineering drawings and drawings output by solid modellers needed properties, tolerances, surface finishes and often dimensioning to be added in the same way as before.

2.2.4 Islands of automation

We have described three islands of automation. Today's problem is to facilitate information flow between the activities and, in many cases, that means automatically on the computer. Most of the processing required to do the conversions between the representations of geometry implicit in the above could be done algorithmically. The difficulty is how to do it without creating new problems or wasting investment in systems and training. An understanding of the form and content of one's product data in its entirety is a prerequisite and for engineering data, in its fullest and widest sense, this is a research topic.

The problem which needed solving first in the 1970s was communication of drawings. Because all drawings were commonly drawn on a CAD system, it was straightforward for the sender to print them out and for the receiver to re-input them. It was realized at an early stage that this need, to communicate drawing information from one partner to another, could be made more efficient. Data exchange technology grew out of this need.

Surfaces came second: early extensions of exchange technology were to enable the exchange of surface representations between the designer and (for example) the tool maker.

The exchange of data for finite-element-based stress analysis has been addressed since the early 1980s, using element/mesh nodes and associated properties. Intelligence here would be geometry with associated properties, which is not yet widely used. There has been a delay in satisfaction with information exchange for finite-element analysis because of this ambivalence between associating attributes with meshes (which is now common) and analyses based directly on the definition geometry with attributes.

2.3 The product life-cycle and suppliers

Different aspects of the product are developed by different people. Consequently, they have different emphases, or even totally disjointed information requirements. Our examples all included some aspect of geometry. The use of engineering drawings is still widespread and they are a common language between life-cycle phases. They are often the most important conveyor of information between contractor, customer and equipment supplier in the process plant and construction industry, between main manufacturer and component supplier, between design consultant and contract manufacturer. All these people do something with the drawing and what they do is usually on a CAD system. This can introduce a new problem. If the CAD systems are from different vendors they may, and usually do, represent the same information in different forms. How would you represent a circle, show that a collection of lines and arcs represents a gear, that text is part of a symbol and changes with the symbol? There are several different levels at which differences can occur. At the simplest level, the same data items can be organized differently. A second is the representation of a dimension by appropriate characters next to the line as compared with being able to calculate the length (and hence the characters and their placement) from geometric information in the model. Even more complex examples occur if we move to other geometric domains. Conversion between surface-representation forms while maintaining accuracy is a new technology and conversion between complete representations of solids is not always possible, even today [125, 126, 124].

So there are two scenarios: islands of automation in a company and dissimilar CAD systems which superficially perform the same function. The first usually has to be solved by the company and will be given priority depending on its perceived influence on profitability. This will require an engineering data management (EDM) or integration solution. The second must be solved to be able to operate in consortia or supplier chains. It can, of course, happen within single companies, especially where mergers have occurred.

There is a variation of the second scenario which may have significant differences – that of archiving. Data often has to last many years. Some aircraft need maintaining for fifty years (DC-3s are still in use), and nuclear plant for longer than that, even if most of its life-cycle is spent in decommissioning. In that time, new computer systems will be in use and (perhaps more significantly) those familiar with the system on which the drawing was generated will have moved away from the enterprise. Until we have confidence that any future computer system can read such data, paper engineering drawings will survive.

We have not described any significant alteration in the design process due to increased use of computers. The applications are aids to allow us to design and make more complicated objects in cost-effective timescales. Neither have we considered the other phases of the product lifecycle, because their effect on today's data exchange scene is minimal (due to the islands of automation). The necessity of imagining a better-managed or more-integrated engineering world is exercising the STEP community and will be discussed later.

2.4 Costs

We have illustrated that there are difficulties in exchanging information between the systems used in engineering and that our uptake of computers has generated new, tedious and error-prone tasks. We have not estimated the cost: this is difficult. Large projects (such as oil platforms and ships) generate hundreds of thousands of drawings on computers which many different parties need to access. Powerful contractors like the US Department of Defence (DoD) or even large car manufacturers are demanding digital data compatible with their systems. The ability to provide this is a matter of survival for suppliers. The amount of time and money that technologically advanced companies and governments are investing in standards for the exchange of product model data (STEP)and their growing number adds to our confidence that the need is real.

2.5 Data exchange

Thus we need to transfer in computer-readable form, product data and particularly product definition data as completely as possible to preserve its information content. With our previous emphasis it is reasonable to say we wish to preserve the design intent. There are several solutions currently being used.

2.5.1 Use of a common system

The most obvious solution is to ensure all parties in collaboration use the same CAD system. This solution is possible on very large projects where perhaps the cost of installing new systems can be hidden. This often happened in large capital projects in the early days of CAD. Many small suppliers with little need to computerize to make their own products believe that this is what the large car manufacturers are saying to them: namely, "you cannot be a supplier unless you buy the same computer system as us". The authors do not believe they are saying that (at least, not any longer), but they may just as well be if cost-effective means of compatibility are not available.

2.5.2 Graphical transfer

A common route has been to transfer the manifestation of the definition data using a graphics format such as the Calcomp format, which became one of the most widely used plotting formats in the early years of CAD. This consists of the line segments of which the picture is composed when it is output on a Calcomp plotter. The format did become quite widespread for transferring output from incremental plotters in the late 1970s. All the design data are lost, but of course a new picture can be generated for the human. There is some value in this, judging by the large number of paper drawings which are now being "scanned" into digital form for computers. They too will suffer the same lack of inherent information content. An advantage is the reduction in storage required when one moves from paper to magnetic form.

2.5.3 Direct translators

A direct translator is a piece of software which takes input data from one system and converts it so that it can be used in a second system to represent the same information if possible. If the sending system can represent a circle explicitly one would not wish to have the circle represented as a conic or a polyline in the receiving system even if the drawing produced looked the same. Direct translation is widely used and has been since the problem of data exchange became apparent in the early 1970s. We consider direct translators to be the first generation of CAD data exchange processors.

Their most common use has been inside large high-technology companies such as those in aerospace who were early in their use of computers and often wrote in-house systems because they were ahead of software houses in understanding the needs of their company and the potential of computers. However, there are still many circumstances where exchange is needed and no other method is available. The geometric-domain transfer of solid geometry and certain special surface forms require special processing to ensure accurate transfer.

Many companies and some software vendors are realizing that geometry is not the only important content of engineering drawings. Information about tolerances and features may be handled intelligently by both systems. Current neutral-format technology does not cope with them, so a direct translator makes sense. There are several software houses that sell direct translators between pairs of the best-known CAD systems, especially where those systems are surface-representation systems. They are apparently thriving on the business. They usually also sell CAD to neutral-format processors and *vice versa*.

The advantages of dedicated translators are:

(a) they may cope where other methods are not available;

(b) there is usually less processing time required, as is generally the case with special-purpose software compared with that designed to cope with a more general situation;

(c) if there is only one pair of systems there is less code to write than when using the neutral format described in Section 2.5.5.

The disadvantages of direct translators are:

(a) The quantity that need to be written increases quadratically with the number of CAD systems between which one wishes to transfer information. Figure 2.2 shows this schematically. The formula for the number of translators required is $n(n-1)$ where n is the number of systems. Thus, for two systems, one needs two processors to transfer both ways; for three systems, six; for four systems, twelve; for five systems, twenty, and so on. Large values of n are not improbable as this includes the case when a CAD vendor updates his system but the old version may still be important. This is compounded if data is being stored for archiving purposes, when it may need to be retrieved by an as-yet-unknown system, possibly many years in the future.

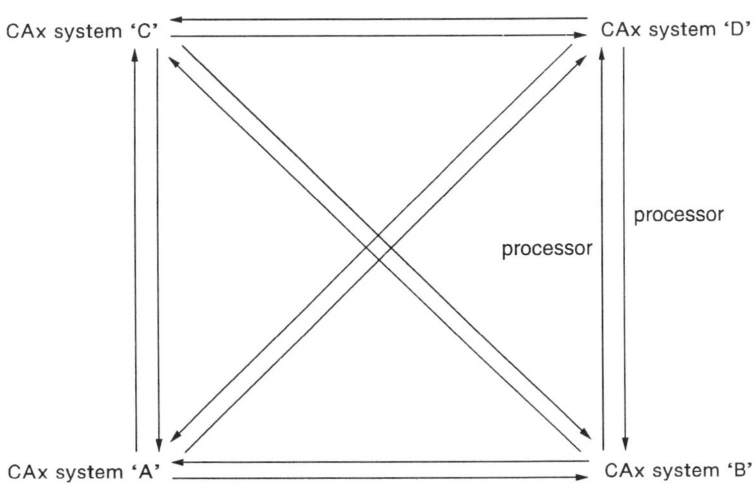

Figure 2.2 Data exchange using direct translators.

(b) The implementor needs to know both systems' representations of the relevant information in some detail. As this is proprietary information the vendor may wish to guard it because he considers it to be of commercial advantage.
(c) There may be doubt over who is responsible for incorrect transfer of information. Is it the implementor of the processor or can he argue that the information the vendor supplied is not adequate? How does one arbitrate with an additional party in the process?
(d) The intelligence and experience that is embodied in a standard or a specification for product data exchange cannot be gained cheaply by any small firm. There is a need to understand the necessary conversions and how to do them.

2.5.4 Vendor formats

Vendor formats are a half-way house between direct translators and neutral formats. The vendor publishes what he calls a neutral format but which is, in reality, closely aligned to his CAD system format. Figure 2.3 shows the neutral-format situation. Any person wishing to transfer to and from that vendor's system has just two processors to write and if the format becomes widely used the situation is in all respects similar to that of using a public neutral format. The vendor obviously guarantees to read any files in his format and can translate his information into the format. However, all the disadvantages with neutral formats are found to occur with few of the advantages. All the formats in this category convey minimal information. They include ISIF by Intergraph, IIF by IBM and DXF by Autocad. The latter's success is largely attributable to Autocad's wide use as an entry-level CAD system and the level of information

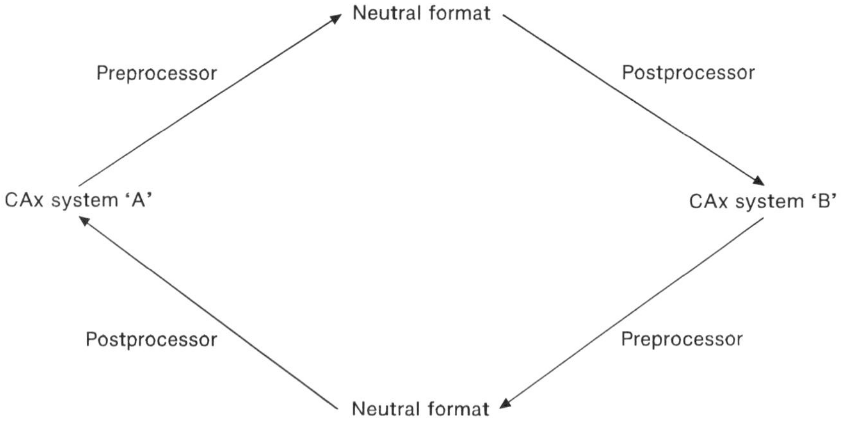

Figure 2.3 Data exchange using a neutral format

content users wish to communicate in current versions. These formats deal in basic geometric elements and some annotation, but little more.

Thus vendor formats are a way of a particular vendor making an anaesthetized version of his internal data structure widely available and thus being able to show that access to his system from any other is easy. The fact that output from his system is similarly easy is not as important to the vendor because it allows users to migrate to other systems. Much of the information that it is possible to store in his system is lost, due to the simplicity of the format he publishes. He does not emphasize this point when trying to encourage a user to buy his system.

2.5.5 Neutral-format file exchange

The method of exchanging information between two dissimilar systems which is theoretically the most effective is called neutral-format file exchange. This is what we term "second-generation" data exchange; it is also called CADCAM data exchange. We differentiate it from third-generation exchange, which will be STEP-based, where the emphasis on CADCAM will disappear. The physical file format as the exchange medium will eventually be just one amongst other implementation methods.

Neutral-format exchange is based on the theory that it is possible to devise a format in which one can represent all information likely to be represented by a CADCAM system in a system-independent form, so that it does not favour any one CAD system representation. Then the vendor of every CAD system can write two pieces of software, commonly called processors, to translate his internal data form into the neutral format and to translate a neutral-format file into his internal data form. Figure 2.3 shows the envisaged situation. A vendor's CAD wares will consist of, at least, his CAD system, a preprocessor (which performs the translation from his system to the neutral format) and a postprocessor (which translates from the neutral format to the vendor's CAD system). The fact that one preprocesses to the neutral file and postprocesses from the neutral file emphasizes that it was the makers of neutral-file formats, largely users with problems, who chose the terminology for data exchange technology, not the CAD system vendors.

The aim of the preprocessor should be to capture all the information content in the CAD system completely. That this may not be possible should be due to real deficiencies in the neutral format being used rather than an incorrect or inadequate processor. The postprocessor must capture the information content of any neutral-format file, retaining as much information content as the system can represent and clearly indicating where information is lost either partially or totally. There are simple CAD systems quite adequate for some uses that do not represent curve forms other than straight lines and circular arcs. Curve-fitting two-dimensional curves by biarcs is a long-standing and well known technique which loses information content of what the original curve was (e. g. a cubic polynomial) but retains a graphic representation.

In the ideal situation above, data exchange would occur as shown in Figure 2.3. Each vendor would have perfect processors and so the user could communicate information from his CAD system to any other and receive information into his CAD system from any other. The only major difficulty would be that CAD systems have differing information contents. That is, of course, what people consider when choosing between CAD systems. Is a two-dimensional draughting system adequate because we just want to facilitate drawing? Are surfaces the crux of our problems? Do we need to use systems with as much information as that in the most progressive and sophisticated geometry systems (e.g. solidity relationships)? This mismatch in functionality of systems even when used in the same phase of the product life-cycle (invariably design), in the same industrial sector and for the same function (e.g. shape definition) is still with us and is the problem with which STEP is grappling. In the 1970s and early 1980s when second-generation data exchange was being developed the major thrust was to exchange the design intent that was currently represented by engineering drawings and their human interpreters.

We have dealt above with the requirements put on the vendors. What do users require of the format? From a functionality point of view it must be independent of any CAD system but complex enough to represent the information content in the most advanced CAD systems. These two needs are somewhat in conflict. The first formats represented well the content of the then current computer-aided draughting systems that enhanced engineering drawing by being able to represent a three-dimensional wireframe model and deal with views and drawing layout, levels and symbols. They could not deal with solids which some systems were starting to incorporate or even surfaces as the most advanced systems represented them. This is not surprising as surface technology is still improving and solid modelling is not yet adequately understood by researchers or users to be the "black box" for use by applications that we had hoped for. So, as with all standards to support information technology, it is easy to denigrate neutral file formats because they are functionally not in line with the state of the art.

A second requirement is that the format be widely supported. It thus has to be as good as possible and become a standard. The process of standardization takes a long time because consensus is essential to ensure acceptance and is an integral part of the procedure. It is tempting to bypass this formal process by publicity about a proposed specification but such action can lead to difficulties with unwilling parties such as vendors or mavericks who have a better idea but do not wish to submit to the rigour of examination by unfriendly peers. There are many examples of *de facto* standards, such as the IBM personal computer with MS-DOS and the various versions of UNIX. The second-generation data-exchange processors did become national standards. International standardization work that is defining STEP led directly from the realization that the community needed an internationally accepted neutral file format to avoid duplication and ever more competing (in the international arena) national standards.

A third thread was not perhaps sufficiently understood when the second-generation formats were first proposed: there must be a mechanism to ensure that processors work correctly. If we want open systems we need standards and for standards to work we need processor testing. This is a large subject which is expanded on in later chapters (Part III). Users must have confidence in the system.

The advantages of neutral format data exchange are listed below.

(a) The number of pieces of software needed is directly proportional to the number of processors involved in the exchanges. Figure 2.3 illustrates this and shows that if there are n systems, $2n$ processors are required. This contrasts favourably with direct translation where the number of processors increases quadratically. As soon as there are more than three CAD systems neutral format data exchange needs less processors to be written.

(b) The responsibility for writing the processors lies clearly with the CAD system vendor. Before purchasing the system, an informed buyer can insist on seeing the reading of a neutral-format file and writing of his information into a file. This is in addition to any other user-acceptance tests. Depending on how important it is to his company, this testing can be a large factor in procurement.

(c) A standard or even a controlled specification can be made the subject of conformance testing. Although this is not a replacement for user-acceptance testing, one will have more confidence in processors that have been conformance tested.

(d) The possibility of standardization means that wide consensus or widespread contribution to the development of the format will occur, ensuring knowledge and experience is captured in the resulting format.

(e) Testing can be undertaken by an independent organization based on an independent document.

The disadvantages of neutral formats are as follows.

(a) It is difficult to find one which has adequate capability as far as information capture is concerned and thus achieve standardization or even *de facto* standardization at a sufficiently high functional capability. In particular, with CAD data exchange neutral file formats, one is attempting to standardize while the technology is still evolving: one does not wish to settle on a solution that will not stand the test of time.

(b) Ensuring that the format is supported by adequate processors is the responsibility of the individual vendors. It is not in their own self-interest to enable easy information flow out of their system, because that allows users to purchase any other CAD system at their next procurement.

Part of the problem with second-generation formats has been inadequate specifications which led to misinterpretation. This and all other technical problems as identified through use of second-generation formats will be described in detail in Chapter 7.

2.6 The standards-making process

Before we go on it is essential that the reader understand the standards-making process. Our experience is that the outsider and those whose main interest is innovation find this far too lengthy and frustrating. Achieving a standard hinges on consensus. People with experience of the field and those with vision are needed. The essential requirement for standards is to make implementation of information technology effective. This suffers, because such a consensus needs widespread understanding, which in such a fast-moving area takes dedicated time from the right people. There is always a good argument to delay backed by the syndrome that standards provide "not equal benefit but equal misery" due to the fact that one tends to reach agreement about the simpler parts. It is often said that a standard may represent the lowest common denominator (strictly, it is the highest common factor).

There are national standards bodies (e.g. BSI, AFNOR, DIN, ANSI), a European body (CEN/CENELEC), and other international bodies (ISO and IEC). Details are given in Appendix B. The methods by which IGES is changed are described in Chapter 7, and the ballot process for STEP in Chapter 11.

In 1985, the ISO subcommittee TC184/SC4 resolved in Paris that none of the second-generation formats provided an adequate basis for product data exchange and that a subgroup should be set up to define the product data exchange standard to replace them all. The initial release of STEP has now been published as an international standard. In Part II, the situation which led to that resolution is described.

2.7 Conclusion

Having considered the above, we hope we have convinced you of the case for neutral-format-based exchange but also indicated clearly the problems which explain why it may currently have a bad name in some places. Solution of the problems involves better understanding of product data in its entirety, of product definition data as generated and used during design, of representation conversion, and of testing to ensure that processors are good enough. With this understanding, second-generation CAD neutral-format exchange can be made to work in the domains for which it was designed and the need to use direct translators will diminish. Although industry cannot wait for STEP in its entirety, it will be designed to be extensible: enterprises can at least start using the well defined geometry in a framework for product data.

Part II

Second-generation methods: the neutral format

The neutral-format approach to product data exchange began at the end of the 1970s with the definition of IGES. Several of the major CAD vendors provided support for it, and user awareness increased following demonstrations at the annual Autofact exhibitions. This part charts the development of the specification, describing an example IGES file and indicating the additional coverage introduced by each version.

There have been problems with the use of IGES. Although these are discussed in more detail in Part III, some are addressed here in order to show how later versions of the IGES specification solved them. The early versions of IGES – and the problems identified with them – spawned other initiatives, both in Europe and the United States. A number of research and development projects were set up to address the longer-term problems. These initiatives (VDA-FS, SET, XBF and PDDI) are described in Chapter 4.

At about the same time as these projects were starting, an international meeting took place which initiated STEP. The history and description of this standard are provided in Part IV, but the reader needs to be aware that much of the development of the existing specifications in the late 1980s both influenced and was influenced by STEP.

The part concludes with an overview of related standards in the field of computer-aided engineering, including those for graphics, networking, documentation and trade data. Some of these, developed independently, provided expertise and methods which were used subsequently by product data exchange standards, whereas others provided an infrastructure into which these standards could fit.

Chapter 3

Early versions of IGES (1979–1983)

3.1 History of development

The Initial Graphics Exchange Specification, known as IGES, can be traced back to 1979 with a meeting at the National Academy of Sciences in the United States. This meeting included representatives from the United States Air Force, Army and Navy, from NASA and from the National Bureau of Standards (NBS), now known as the National Institute of Standards and Technology (NIST).

The technical committee that was formed to develop IGES was part of a project in the US Air Force ICAM program. NBS was selected to direct and coordinate the IGES work. The participation of Boeing and General Electric was voluntary and not funded by ICAM. The basis for IGES included the Boeing CIIN format and the General Electric Neutral Database format as well as other formats that had been drafted at the request of the ANSI Y14 Committee. ANSI had responded in the early 1970s to the need to exchange computer-based draughting information, specifically that on a blueprint, by asking for voluntary contributions. By January 1980, IGES Version 1.0[1] was published, containing geometry, graphical elements and annotation for blueprints [75]. The structure of a CAD database could also be captured. In addition, a committee was set up to maintain IGES, to ensure adoption by the technical community, and to provide for extension. The acronym IGES is almost synonymous with CAD data exchange using a neutral-format file.

During the 1970s, the need to exchange engineering drawings in computer-readable form was real. In parallel to the ANSI work, *ad hoc* solutions were appearing including vendor formats (e.g. IIF from IBM and ISIF from Intergraph). More academic work was promoted by the Geometric Modelling Project of CAM-I [24] which, rather surprisingly, concentrated on a format to facilitate the exchange of solid models. This was led by the need for representation conversion between solid forms in order to provide solid modelling systems which could support enough applications (i.e. had enough functionality) to make them usable. Because of this lack of comprehensive systems, solid modelling exchange has only now reached the level of use where a standard is required. The early work by CAM-I was executed by McAuto who based their work on the ANSI initiative and produced formats for constructive solid geometry and boundary representation.

In Spring 1980, ANSI Y14 adopted IGES 1.0 and the McAuto work as the proposed standard: the ANSI Y14.26M document appeared in January 1981 in five parts [76]. The first four parts were IGES 1.0 and the fifth part was the McAuto work of CAM-I.

Before starting a description of IGES, its development up to Version 2.0 is recorded. The early momentum was continued and committees were established to add to IGES to cater for the requirements of electrical schematics, finite-element analyses and plant design. Plant design under the AEC Committee was delayed to Version 3.0. IGES Version 2.0 was released in 1983 but was not submitted to ANSI for standardization. It was used until IGES 3.0 appeared in 1986. Delays were starting to slow progress largely because less well understood concepts were being addressed and perhaps because those actively interested were considering the next generation of data exchange, product data exchange.

The McAuto file specification did not satisfy the CAM-I requirement. Instead, the Geometric Modelling Project of CAM-I in 1982 developed XBF [173] which is a direct forebear of current solid modelling representations in IGES and STEP.

Figure 0.1 shows product data exchange evolution from 1980 onwards, in which the place of these early contributions is included.

3.2 IGES 1.0

3.2.1 Description

As far as is known, the IGES initiative was the first to attempt to define a specification that would be independent of CAD system data formats. The intention was to create a specification that could be used for the "communication of product definition data created and used by CADCAM systems"; that is, the information "required to describe and communicate engineering characteristics of physical objects as manufactured products". In the preamble to the IGES specifications, the phrase "design intent" is used to stress a level of information content higher than that which graphical representation alone can transfer. In the 1970s, this meant the functionality of that generation of CAD systems which allowed one of two types of shape definition. Such systems used either the traditional orthographic projections augmented by wireframe representations in three dimensions or allowed the representation of surfaces by some form of polynomial patch system. Entities could also be defined in a two-dimensional plane and then placed in a three-dimensional environment by a transformation matrix, allowing the transfer of so-called 2½-dimensional information.

It should be mentioned that IGES first addressed mechanical CAD primarily. IPC was a concurrent 1970s initiative from the electrical industry (see Ch. 6).

3.2.2 Functionality

IGES 1.0 [75] covers geometries which may have coordinates in two-dimensional or three-dimensional space. The three-dimensional co-ordinates are used to represent wireframe objects which may have planar or curved surfaces. The specification also covers annotations including a range of common dimensioning formats. There is also a collection of structural facilities which allow relationships (such as groups) and attached properties. This functionality is usually shown schematically as in Figure 3.1; thus, an IGES file is composed of entities which are:

(a) geometry,
(b) annotation, or
(c) structure.

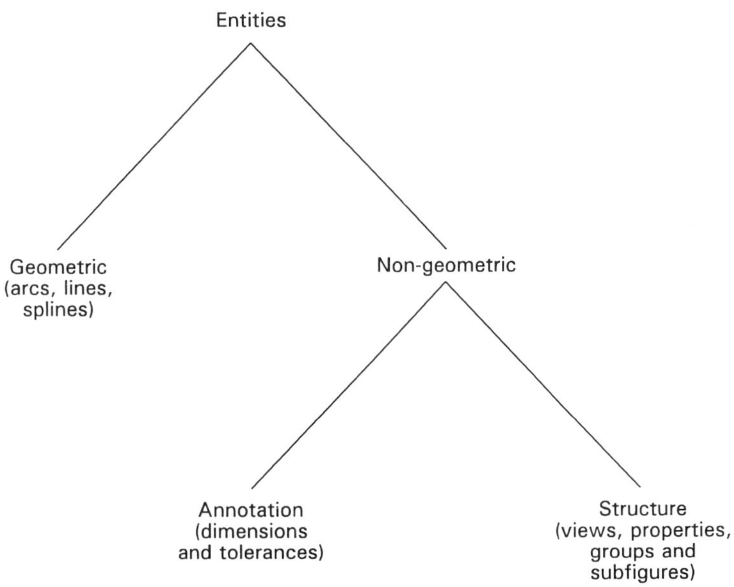

Figure 3.1 IGES entity structure.

In addition, provision was made for associativity and macro definitions, so that implementors could define entities for constructs in their system which it was not possible to capture in IGES directly. A receiving system should thus have been able to retain the structure originating from such a sending system, even if it could not "understand" the details: it would merely store the data and incorporate it back into the IGES file when it, in turn, became the sending system.

In order to use IGES, the relevant entities were written to an ASCII file made up of 80-character records which, in the days of IGES 1.0 and 2.0, was then written to magnetic tape.

3.2.3 File format

A sample IGES file is shown in Figure 3.2. One can see by examination of that file that the IGES 1.0 file format has five sections identified by a character in every column 73 (the 73rd character position) as follows:

S: start,
G: global,
D: directory entry,
P: parameter data,
T: terminate.

Each record has a sequence number in columns (character positions) 74 to 80 which indicates its position in the section to which it belongs. The legacy of Fortran and the punched card is glaringly obvious.

Start section

This section is intended for human readership. It is an unformatted prologue to the file in which the sender can put messages to the receiver about agreed conventions, such as the use of layers and what is contained in each layer. There are no guidelines in IGES 1.0 on how this section should be used.

Global section

This section is a single logical record which may extend over several physical records of eighty characters. Each has a G in position 73 and sequence number in positions 74 to 80 of each line as described above. There are twenty-two parameters in this section, in free format. This means that the parameters are separated by delimiters and the record is terminated by a delimiter; the "columns" in each record are not significant, apart from those reserved for the sequence number. The characters chosen for these delimiters are defined in this section and in our example are the default values: "1H," to separate the parameters and "1H;" to terminate the record.

```
 Example IGES file for demonstration purposes only                       S      1
1H,,1H;, 12HExample IGES,   7HEX.IGES,   5HLuigi,  11Hversion 4.0,       G      1
 32,  38,   6, 308,  15,  12HExample IGES,  1.00000E+00,   2,   2HMM,    G      2
     1,  0.00000E+00,  13H920819.000000,  1.00000E-06,  8.00000E+01,     G      3
  8HJon Owen,  27HThe University of Leeds, UK,   6,   0;                 G      4
     110       1       0       0      63       0       0       000000000D      1
     110       0       0       2       0                   line        1D      2
     100       3       0       0      63       0       0       000000000D      3
     100       0       0       3       0               circ arc        1D      4
     104       6       0       0      63       0       0       000000000D      5
     104       0       0       5       1               conic           1D      6
     102      11       0       0      63       0       0       000000000D      7
     102       0       0       2       0               compcurv        1D      8
     406      13       0       0      63       0       0       000000300D      9
     406       0       0       2       1               deflevel        1D     10
     110      15       0       1      -9       0       0       000000000D     11
     110       0       0       2       0                   line        2D     12
     110      17       0       1      -9       0       0       000000000D     13
     110       0       0       2       0                   line        3D     14
     110      19       0       1      -9       0       0       000000000D     15
     110       0       0       2       0                   line        4D     16
     124      21       0       0      63       0       0       000000000D     17
     124       0       0       4       0                 matrix        1D     18
     410      25       0       0      63       0      17       000000100D     19
     410       0       0       1       0                   view        1D     20
     410      26       0       0      63       0       0       000000100D     21
     410       0       0       1       0                   view        2D     22
     410      27       0       0      63       0       0       000000100D     23
     410       0       0       1       0                   view        3D     24
     404      28       0       0      63       0       0       000000100D     25
     404       0       0       5       0                drawing        1D     26
     402      33       0       0      63       0       0       000000300D     27
     402       0       0       2       7                  group        1D     28
     110,  1.00000E+00,  2.00000E+00,  3.00000E+00,                    1P      1
           4.00000E+00,  5.00000E+00,  6.00000E+00;                    1P      2
     100, -1.00000E+00, -2.50000E+00, -2.50000E+00,                    3P      3
                        -3.25000E+00,  4.60000E+00,                    3P      4
                        -3.25000E+00, -4.00000E-01;                    3P      5
     104,  1.00000E+00,  2.00000E+00,  3.00000E+00,                    5P      6
           4.00000E+00,  5.00000E+00,  6.00000E+00,                    5P      7
          -1.00000E+00,                                                5P      8
           7.00000E+00,  8.00000E+00,                                  5P      9
           9.00000E+00,  1.00000E+01;                                  5P     10
     102,          3,                                                  7P     11
                   1,           3,            5;                       7P     12
     406,          4,                                                  9P     13
                  60,          61,           62,          63;          9P     14
     110,  0.00000E+00,  0.00000E+00,  0.00000E+00,                   11P     15
           1.00000E+00,  0.00000E+00,  0.00000E+00;                   11P     16
     110,  0.00000E+00,  0.00000E+00,  0.00000E+00,                   13P     17
           0.00000E+00,  1.00000E+00,  0.00000E+00;                   13P     18
     110,  0.00000E+00,  0.00000E+00,  0.00000E+00,                   15P     19
           0.00000E+00,  0.00000E+00,  1.00000E+00;                   15P     20
     124,                                                             17P     21
         1.00000E+00,  0.00000E+00,  0.00000E+00, -3.00000E+00,       17P     22
         0.00000E+00,  1.00000E+00,  0.00000E+00, -4.00000E+00,       17P     23
         0.00000E+00,  0.00000E+00, -1.00000E+00, -5.00000E+00;       17P     24
     410,         42,  1.00000E+00;                                   19P     25
     410,         43,  5.00000E-01;                                   21P     26
     410,         44,  2.54000E+01;                                   23P     27
     404,          3,                                                 25P     28
                  19,  1.00000E-01,  2.00000E-01,                     25P     29
                  21,  3.00000E-01,  4.00000E-01,                     25P     30
                  23,  5.00000E-01,  6.00000E-01,                     25P     31
                   0;                                                 25P     32
     402,          4,                                                 27P     33
                   1,          11,           13,          15;         27P     34
S      1G      4D     28P     34                                       T      1
```

Figure 3.2 Example of an IGES file.

Table 3.1 IGES global section fields.

Number	Description
1	parameter delimiter (string)
2	record delimiter (string)
3	product identification from sending system (string)
4	file name (string)
5	system identification (string)
6	preprocessor version (string)
7	number of bits for integer representation (integer)
8	maximum power of ten in single-precision (integer)
9	number of significant digits in single-precision (integer)
10	maximum power of ten in double-precision (integer)
11	number of significant digits in double-precision (integer)
12	product identification for receiving system (string)
13	model-space scale (real)
14	unit flag (integer)
15	units (string)
16	maximum number of line-weight gradations (integer)
17	maximum line width (real)
18	date stamp of file generation (string)
19	minimum user-intended resolution (real)
20	approximate maximum value (real)
21	author (string)
22	organization (string)
23	IGES version (integer)
24	drafting standard (integer)
25	date stamp of model (string)

In IGES, fields 8 and 9 contained the value of the number of bits in a single-precision floating point exponent and mantissa, respectively. Fields 10 and 11 held the same values for double-precision. Fields 23 and 24 were added in IGES 3.0; field 25 in IGES 5.0.

The parameters contain information about the CAD model and the system that generated it, including model identifiers, modelling scale, units and the size of integers and real numbers in the sending system. They are listed in Table 3.1.

In both the global section and the parameter data section, text data is represented as a Hollerith string. The form of a Hollerith string is "nHcharacterstring", where "n" is the number of characters in the string, "H" is the opening delimiter and "characterstring" is any collection of ASCII characters. An example is 9HIGES DATA. This is another Fortran legacy which dates IGES.

Directory entry section

This section is effectively a table of contents: every entity in the file has a logical record of fixed format, comprising two physical lines each

1 entity type number	2 parameter data pointer	3 (structure)[1]	4 line font	5 level	6 view	7 matrix	8 label display	9 status number[2]	10 sequence number
11 entity type number	12 line weight	13 colour[3]	14 parameter data line count	15 form	16 (reserved)	17 (reserved)	18 entity label	19 entity subscript	20 sequence number

Figure 3.3 IGES directory entry fields.

Notes:
1 Was version in IGES 1.0 and 2.0
2 Blank, subordinate, entity use, hierarchy
3 Was pen number in IGES 1.0

containing ten fields with eight characters in each field. The fields are shown in Figure 3.3.

Parameter data section

Parameter data – such as coordinates, coefficients and relationships – vary from entity to entity. A full description is contained in the IGES documents; the names and types of the attributes of each entity are given in Appendix E. The data in this section are also in free format. A parameter data record may consist of several lines and each entity specified in the directory section must have at least one parameter data line. The definition of the entity in the IGES specification gives the order of the attributes in the parameter data section. Table 3.2 shows those attributes which are specific for a circular arc, entity 100, as presented in IGES 1.0. (After the entity-specific information, each entity may include back-pointers to associativities and properties.) Character positions 1 to 64 contain the parameter data in free format. The first entry is always the entity type number and its value must be the same as that in fields 1 and 11 of the corresponding directory entry record. Character position 65 is always blank; 66 to 72 contain a pointer to the directory entry (its line number in columns 74 to 80 of that section); 73 to 80 are filled as described above.

Table 3.2 Parameter data for IGES 1.0 circular arc entity.

Index	Name	Type	Description
1	ZT	real	parallel ZT displacement of arc from XT, YT plane
2	X1	real	arc center abscissa
3	Y1	real	arc center ordinate
4	X2	real	start point abscissa
5	Y2	real	start point ordinate
6	X3	real	terminate point abscissa
7	Y3	real	terminate point ordinate

Terminate section
This is a single record which contains totals of the number of records in each of the start, global, directory entry and parameter data sections. It is used as a checksum.

Because the global and parameter data sections are in free format, some integer and real values may be omitted (shown either by the presence of two consecutive parameter delimiters or by a record delimiter earlier than expected). In these cases, the default value of zero is assumed. However, in certain cases (such as for angles), an entity-specific default applies (in the case of slant angles for text, this is usually $\pi/2$). Further details are in Appendix E, but the IGES specifications should be consulted for full details.

3.2.4 The entities

This section contains an explanation of the sample IGES file shown in Figure 3.3. It is not intended to describe all the IGES entities fully but to choose a few to indicate how they are specified and to show the complex relationships for which IGES 1.0 catered. The complete list of IGES 1.0 entities is shown in Table 3.3.

The IGES file shown in Figure 3.3 contains instances of the following entities:

110 line;
100 circular arc;
102 composite curve;
104 conic arc;
124 transformation matrix;
410 view;
404 drawing;
406/1 property – definition levels;
402/7 associativity – group (without backpointers).

First, consider the parameter data, denoted by a "P" in column 73. The line has two endpoints specified in three dimensions: note that the IGES file has been laid out to facilitate human readability. The circular arc has a z-displacement, followed by centre, start and endpoints in two dimensions. The conic has the six coefficients of its equation, followed by its z-displacement, and start- and endpoints, again in two dimensions. These geometric entities may now be combined into a composite curve, the parameter data for which indicates that there are three constituents, which are defined on directory entry lines 1, 3 and 5. In the parameter data for the transformation matrix, the first three reals on each line specify a rotation, and the fourth a translation.

The directory entry of these entities may now be considered. Each entry comprises two physical records, the first field of each of which is

Table 3.3 IGES 1.0 entities.

Entity	Form	Description
100		circular arc
102		composite curve
104		conic arc
	0	determined by parameters
	1	elliptic
	2	hyperbolic
	3	parabolic
106		copious date
	1	data pairs in a plane
	2	data triples in a plane
	3	data sextuples in a plane
	11	2D linear path
	12	3D linear path
	13	piecewise linear string
	20	centreline through points
	21	centreline through circle centres
	31	section: solid
	32	section: solid pairs
	33	section: solid + dash
	34	section: solid quads
	35	section: solid + dash + solid
	36	section: dash
	37	section: solid + perpendicular solid
	38	section: solid + perpendicular dash
	40	witness line
108		plane
	-1	bounded void
	1	bounded
110		line
112		parametric spline curve
114		parametric spline surface
116		point
118		ruled surface
	0	equal relative arc length
120		surface of revolution
122		tabulated cylinder
124		transformation matrix
	0	orthonormal rotate/translate
202		angular dimension
206		diameter dimension
208		flag note
210		general label
212		general note
	0	simple note (default)
214		leader (arrow)
	1	wedge
	2	triangle
	3	filled triangle
	4	no arrowhead
	5	circle
	6	filled circle
	7	rectangle
	8	filled rectangle
	9	slash
	10	integral sign
216		linear dimension
	0	undetermined form
218		ordinate dimension
	0	witness line or leader
220		point dimension
222		radius dimension
	0	single leader format
302		associativity definition
304		line font definition
	1	subfigure pointer
	2	repeating pattern
306		macro definition
308		subfigure definition
310		text font definition
402		associativity instance
	1	unordered group with backpointers
	3	views visible
	4	views visible, color, line weight
	5	entity label display
	6	view list
	7	unordered group without backpointers
	5001..9999	implementor defined (302 required)
404		drawing
	0	translate and scale
406		property
	1	definition levels
	5001..9999	implementor defined
408		singular subfigure instance
410		view
	0	orthographic parallel
600..699		implementor defined macro instance

the entity number, to match that at the start of the corresponding parameter data record. Fields 18 and 19 have been used to indicate the name of each entity type and its number in the file (view 1, view 2, conic 1, . . .). All entities have their colour (field 13) set to zero, which indicates that it is undefined (and hence any may be used). Similarly, all entities have their line font (field 4) also set to zero, apart from the group of three lines which will be presented explicitly as solid lines (value of one).

The file pointer structure is realized using field 2 of each directory entry, which points forward to the parameter data. Field 14 contains the number of lines in the corresponding parameter data record. (Note that this facility can be used by a receiving postprocessor to ignore the correct number of parameter data lines of an entity which is not supported.) Each parameter data record contains a backpointer to its corresponding directory entry record in columns 66 to 72, as noted earlier. There is no requirement to have the entities in the same order in the directory entry and parameter data sections, but in this case they have been ordered in this way. As a result, if the values in fields 2 and 14 of a directory entry record are added, the value in field 2 of the next directory entry record is the total.

The only geometric entity to have a form number (field 15 in the directory entry) is the conic, in which it denotes an elliptic arc (form 1); both the property (form 1) and the associativity (form 7) also have a non-zero value for the form.

More complex structures can now be considered. The single drawing, in its parameter data, contains three views, each of which is defined by a pointer to its directory entry and an origin in two dimensions. (Annotation may also be placed on a drawing, but has not been in this example.) Each view has a view number and a scale factor. (Clipping planes may be specified, but have not been included in this example.) However, each view does not specify explicitly which entities appear in it. Instead, field 6 of the directory entry of each entity indicates in which view it belongs.

The first few entities in this example appear on level 63, as shown by field 5 of the directory entry records. However, using the property entity, some entities are designated as appearing on more than one level. Each of the group of three lines is placed on levels 60 to 63: field 5 of the directory entry points to the directory entry of the definition levels property, whose parameter data indicates a set of four levels with these numbers. An instance of this property is required for each set of levels on which entities appear.

All of the lines in the example have been grouped together: the parameter data of the associativity indicates the four entities defined by the directory entry record numbers which follow.

Finally, all geometric entities are untransformed; that is, they are specified in world co-ordinates. However, using field 7 of the directory entry of the first view, all entities are transformed by the matrix which this field indicates. It is possible to specify sequences of transformations, by placing entities in their own local coordinate system, moving them to that of another "parent" entity, and so on, but this example does not show this feature.

This example is artificial in that it is small, contains invalid values for some of the parameters and has been kept as simple as possible, but it is representative in that the entities described convey engineering drawing data, and continue to do so today.

3.3 IGES 2.0

IGES Version 2.0 [77] is a refinement and extension of Version 1.0 as indicated in the introduction to this chapter. IGES 2.0 started the improvement of the documentation which is still continuing. The major change

Table 3.4 New entities and forms in IGES 2.0.

Entity	Form	Description
106		copious data
	63	simple closed planar curve
118		ruled surface
	1	equal relative parametric values
124		transformation matrix
	10	cartesian coordinates
	11	cylindrical coordinates
	12	spherical coordinates
125		flash
	0	defined by referenced entity
	1	circular
	2	rectangle
	3	doughnut
	4	canoe
126		rational B-spline curve
	0	determined by parameters
	1	line
	2	circular arc
	3	elliptical arc
	4	parabolic arc
	5	hyperbolic arc
128		rational B-spline surface
	0	determined by parameters
	1	plane
	2	right circular cylinder
	3	cone
	4	sphere
	5	torus
	6	surface of revolution
	7	tabulated cylinder
	8	ruled surface
	9	general quadric surface
134		node
136		finite element
402		associativity instance
	8	signal string
	9	single parent
	10	text node
	11	connect node
406		property
	2	region restriction
	3	level function
	4	region fill
	5	line widening
	6	drilled hole
	7	reference designator
	8	pin number
	9	part number
	10	hierarchy
412		rectangular array subfigure instance
414		circular array subfigure instance

was to existing entities and the inclusion of new entities, which are listed in Table 3.4. They can be summarized as follows:

(a) two new geometry entries: rational B-spline curve and surface;
(b) enhancement of ruled surface, tabulated cylinder and surface of revolution entities;
(c) larger set of character fonts;
(d) provision for some finite-element modelling data;
(e) electronics printed wiring board data.

IGES 1.0 and IGES 2.0 are upward compatible except that there is a change in the text font definition entity (310). Because of complaints about IGES file size (due to the 80-column card format, the fixed-format records and the comprehensive cross-checking data) a binary file format was introduced. That the file is written in binary was indicated by an optional section before the start section which has a B (for binary) in character position 73. IGES 2.0 was never submitted for standardization, possibly because of the time this would take and optimism about the early appearance of IGES 3.0, although most early implementations were based on IGES 2.0 rather than IGES 1.0.

3.4 ANSI Y14.26M

Vendors used IGES 1.0 and 2.0 rather than the ANSI standard document [76]. However, the ANSI document was more readable as it resolved some of the layout and nomenclature inconsistencies in IGES 1.0. It also included provision for solid modelling in section 5.0, entitled "basic shape description". Three-dimensional models are built using a "bottom-up" approach. There are three types of transformations:

(a) interpolative (linear, circular/conic, spline, *n*th-degree polynomial);
(b) generative (translate and/or rotate); and
(c) replicative (translate, rotate, scale, reflect).

Using these, structures and points may be made into lines, surfaces and volumes. Each structure adds a dimension onto the entity upon which it operates. Point set operators (union, difference and intersection) may then be applied to the volumes. It is also possible to specify the topology of an object in a similar manner: vertex into edge, loop, face, shell and, finally, object. Although CAM-I rejected the work, it made interesting reading to those with a working interest in solid representation.

Figure 3.4 The 1983 and 1993 committee structures of the IGES/PDES Organization.

3.5 The early promotion of IGES

3.5.1 The IGES committees

IGES was and still is supported by a committee structure [163], as shown in Figure 3.4. Under the leadership of NBS, a program to inform managers, technical users in vendor companies and end-user companies was started.

3.5.2 Autofact

At the annual Autofact exhibition, conference and demonstrations held in 1980, three major CAD vendors (Applicon, Calma and Computervision) announced an initial capability for IGES to which Control Data Corporation added their name in 1981. At the National Computer Graphics Association (NCGA) meeting in 1982, the above four vendors plus Gerber, McAuto-Unigraphics and Manufacturing and Consulting Services (MCS) all either provided tapes or demonstrated an IGES capability and, by Autofact 1983, 32 vendor companies had indicated commitment to IGES [142].

A test library was produced by the IGES Organization in autumn 1981 which was intended to facilitate testing. In 1981 the three-dimensional model shown in Figure 3.5 was transferred from a Computervision CADDS4 system in Sandia's National Laboratory at Livermore, California

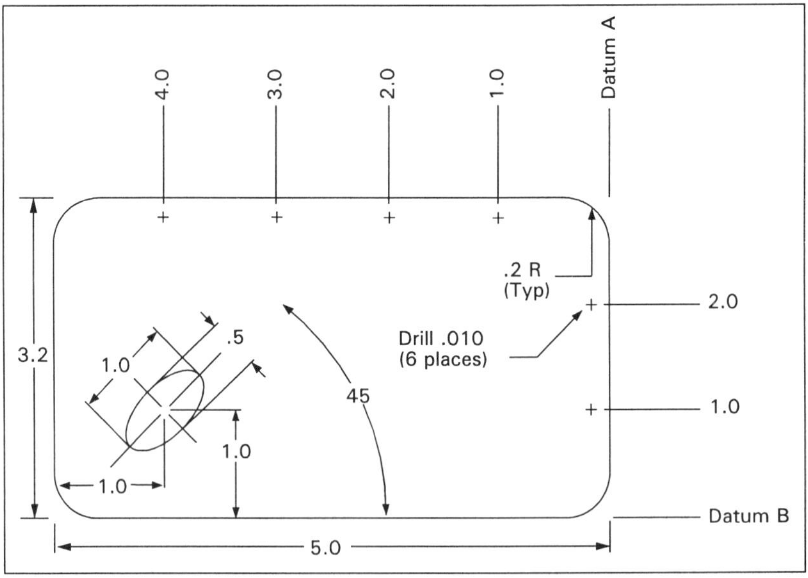

Figure 3.5 Part used in the 1981 Autofact demonstration.

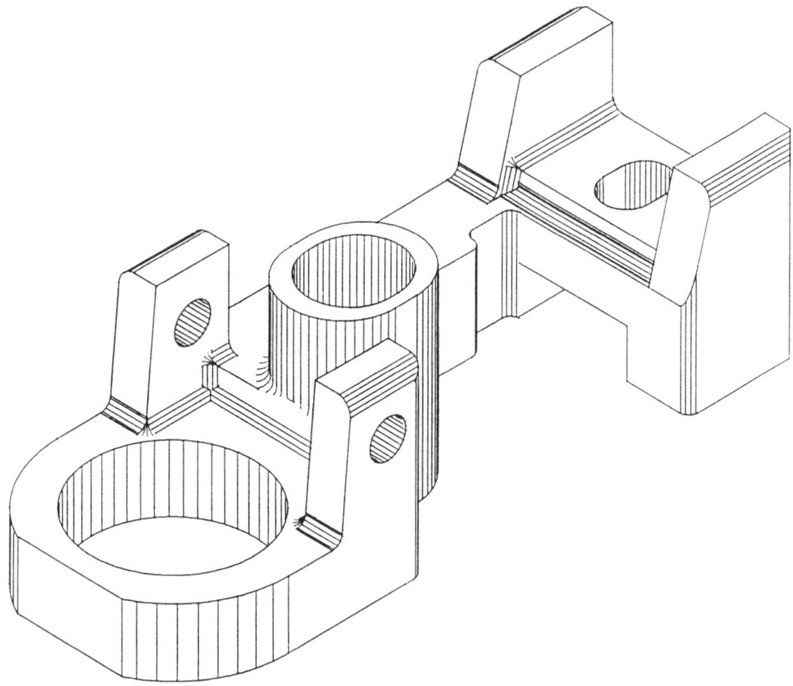

Figure 3.6 MBB part, used in the Autofact 1983 demonstration.

to a CDC2000 system at the Bendix Corporation in Kansas. A public test was arranged for the 1982 NCGA meeting at Anaheim in which Applicon, Calma, Computervision, Gerber and MCS participated. The major problems were reported to be caused by magnetic tape incompatibilities and with arc placements. In December 1982, at Autofact in Philadelphia, Computervision, CDC, Gerber, IBM and MCSI (as MCS had become) staged a similar demonstration. The ubiquitous MBB part, shown in Figure 3.6 without the accompanying annotation and dimensions, was the test part at Autofact 5 in Chicago in December 1983 at which twelve vendors agreed to participate. In view of what we say later it must be remembered that these demonstrations were stage-managed. Those of us with experience of demonstrating will know how wise this is, but such demonstrations indicate little about the performance of IGES implementations for exchange when used without such careful preparation.

3.5.3 Outside the United States

These demonstration and testing activities were understandably focused in the United States and an early complaint in Britain was the difficulty of obtaining information. The obvious source was one's CAD vendor but, because many were American companies and research was centred in the United States, contacts in Europe and elsewhere were either not informed or were given the message that IGES did not work.

3.6 Problems with IGES 1.0 and 2.0

At that time (1983), very few people were able to exchange computer-aided draughting information using IGES. Most of those who were trying were doing test runs and attempting to understand the problems. In this section we shall restrict ourselves to problems inherent in IGES 1.0 and 2.0. Chapter 7 in Part III covers the general problems of data exchange.

3.6.1 Potential problem areas

Terminology

Each CAx system has its own terminology, although the meaning for certain terms may vary from one CAx system to another. For example, one system may store all data on a single drawing, which is composed from several projected views of a model, and accompanying annotation. These views and annotation may appear on one or more levels, each level corresponding to a single piece of paper or Mylar in the draughting office. Alternatively, each piece of paper may correspond to a drawing in its own right. Entities which have some logical grouping may be placed on a single level, or grouped according to some associativity in the CAx system.

CAD system characteristics

In some systems, particularly those for plant design and solid modelling, there is no mechanism for storing two-dimensional information *per se*: all geometry is stored as a three-dimensional model. Such a system cannot sensibly receive information for two dimensions and store it in data structures for three dimensions.

In other systems, although the designer can use the system strictly for two-dimensional draughting, he may prefer to input his model in three dimensions. One or more two-dimensional projections are the means of viewing the model defined in three dimensions. A purely two-dimensional system cannot deal with this situation.

Most IGES geometric entities are specified in three dimensions; that is, they have *x*-, *y*- and *z*-co-ordinates. Two-dimensional systems translate into IGES entities using a constant *z*-co-ordinate. In order to receive a model defined in three dimensions, the projections onto two dimensions must be performed, either by the sending system (preferably) or the receiving system. It is unclear how the receiving system could perform these transformations.

In these cases, the structure of the information behind the drawing will have been lost.

Levels

Every IGES entity may belong to a level, but the maximum number of levels available in CAD systems varies widely. Consequently, a receiving system which has fewer levels than the input file contains needs to map the levels into a smaller number of levels. The IGES start section could be used to convey the mapping from a many-level system to a few-level system, but this would be a manual operation because the start section is human-readable rather than machine-readable. "Closing the loop" (when the drawing is transmitted back to the original system) will necessarily result in the many-level structure having been lost.

Symbols and libraries

In some systems, symbols can be parameterized in a more flexible manner (e.g. shearing) than IGES subfigures allow (by size, position and orientation). The only IGES facility for this is the Macro language and this syntax may not map well onto the sender or receiver system (indeed it may require considerable effort to effect the transformation).

In other systems, symbols are stored by reference to a library, so that when the library symbol is updated any future reproduction of a drawing containing a reference to the symbol is updated similarly. This may clearly be useful to maintain company standards, but such updating may prevent the exact reproduction of an archived drawing. In an IGES environment, there must be a "run-time" definition of the symbol. IGES 2.0 files were designed to be self-contained and so the requirement to pass a drawing referring to a library was outside the scope of IGES 2.0.

If users wished to obtain information about component make-up of a drawing by referring to a standard component/cell library, the library could be passed as a separate IGES file and manually installed in the receiving system.

Size and scale

CAD system handling of model size and plotting scale varies widely. There are basically two approaches. In some CAD systems, the user defines a model (in either two or three dimensions) in terms of real sizes which the system stores. The model may then be viewed by zooming or by applying viewing boxes, which allow the user to expand detailed areas and define what is to be plotted and at what size and scale. The data is sent using IGES as a model and views, which the receiver then scales as appropriate to plot or view the data.

Other systems require the user to know the plotter size and scale accordingly.

Although the first approach is preferred, the other makes the handling of graphical attributes (such as character sizes and line thicknesses) more straightforward: these are sized to make them readable rather than in relation to a particular viewing scale in the drawing. A mixture of these two approaches may be used in yet other systems.

It is advisable that all parts of a drawing/model be to the same scale. One of the most difficult areas is the correct transfer of the sizing, spacing and positioning of characters, and a constant scale removes one area which causes these difficulties.

Connectivity
The concept of connectivity is of fundamental importance to the electronic and process engineering industries, but is captured in different ways by different CAD systems. Some systems enable a "logical" connection to be made between entities, whereas others rely on certain geometric attributes being separated by a particular physical distance.

The major problems concerned in the transfer of data using early versions of IGES appeared to be not at the geometric level (where the mappings are relatively well defined) but at a higher, logical structure level. This required the mapping of complex data structures without degeneration, because any loss of information is accompanied by the overhead of manual editing and the generation of large quantities of data.

It was suggested [16] that a minimum of changes were made by the IGES preprocessor in the sender's data and all accommodation be a result of the postprocessor interpreting what it receives as best it can. Thus, the postprocessor should be a complete implementation of IGES: a subset will not do.

3.6.2 Actual problems with IGES

The classes of problems identified with early IGES transfers were:

- (a) poor preprocessors and postprocessors;
- (b) user misunderstanding of the planning needed to effect a successful transfer;
- (c) inability to read magnetic tapes;
- (d) the size of the IGES files and the time it took to process them;
- (e) inability to translate all the information stored in the sending CAD system.

Of these problems only the size of the files and some of the reasons for poor processors are directly attributable to IGES. Of course, if one wished to translate information for which IGES had no capability, that too would be the fault of the specification. The amazing number of difficulties found in reading magnetic tape and the need to have cooperating parties who discussed procedures to be used in the drawing office and on the CAD system arose because of the lack of experience of the end-users and the fact that IGES was their initiation into exchanging computer-readable material.

The inadequacy of the processors had several causes. The early IGES documents were very difficult to read. Their layout was confusing and off-putting. In addition, the specification was written in a natural language (English) and that led to ambiguities so that anyone implementing IGES had to make judgements which were likely to be different from the next person's and influenced by the particular CAD system for which the processor was being written. An example of ambiguity is found in the definitions for the plane and point entities (108 and 116 respectively): the plane has a "display symbol", whereas the point has a "subfigure instance specifying the display symbol". (In fact, IGES 5.0 changed the latter to a subfigure definition anyway.)

In addition, IGES did not define what constituted an IGES implementation so vendors implemented a small set which matched their CAD system entities. Often, these allowed only graphic representations to be transferred and not any underlying structure. This meant that in any exchange the chances that entities matched and were translated completely were small. That the hearts of vendors were not in facilitating complete transfer, in general, did not help IGES.

Figure 3.7, reproduced from Reference 107, shows an example of a simple drawing. This is used to illustrate the things that can go wrong when IGES is used to transfer drawings. Some examples of the problems encountered are provided in Figures 3.7 to 3.11. When the drawing is plotted which corresponds to the IGES file received (see Fig. 3.8), the problem in Figure 3.7 is clear: all annotation has been lost. Consequently, no dimensioning information is available. However, it may be that the information is actually in the IGES file, but on a level number which is outside the range of the receiving system (and so ignored by it). Equally, such information may have had its attributes set to be subordinate to the drawing, which in turn may not have been processed by the receiving system. Alternatively, the text may be held in a different co-ordinate system. The transfer may not be as catastrophic as it may first appear.

The problem in Figure 3.9 is not visually as obvious: some of the centrelines have become solid. Although this can be interpreted fairly easily by a human reading the drawing, the effect of this change on downstream automation – such as a program which generates numericallly controlled (NC) cutter paths – may be disastrous. A problem may also occur with geometry whose visual appearance is acceptable, but in which some small gaps – due to rounding error, numerical calculation, or even number representation – are present. This, too, would make the

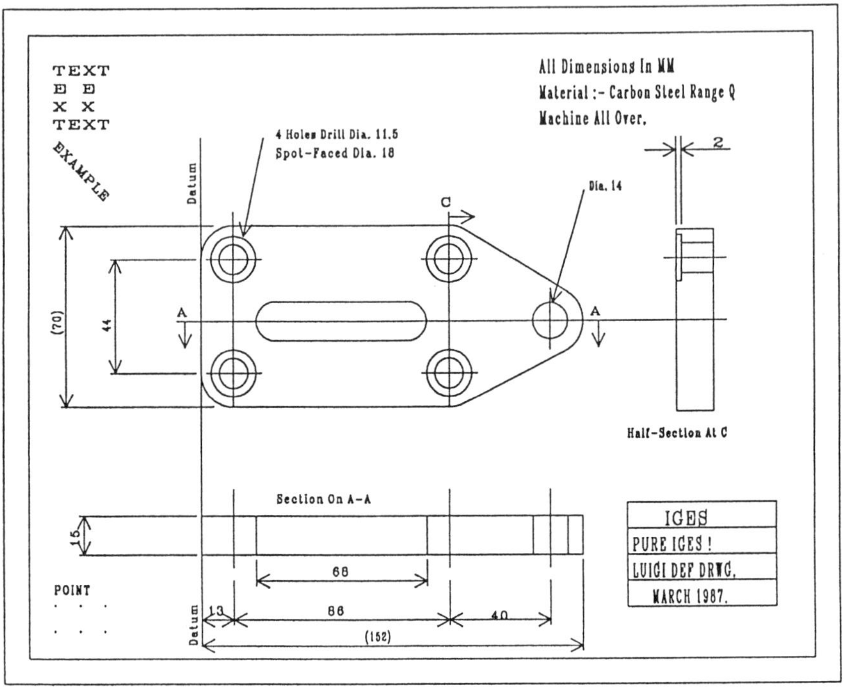

Figure 3.7 Original drawing reproduced from example IGES file.

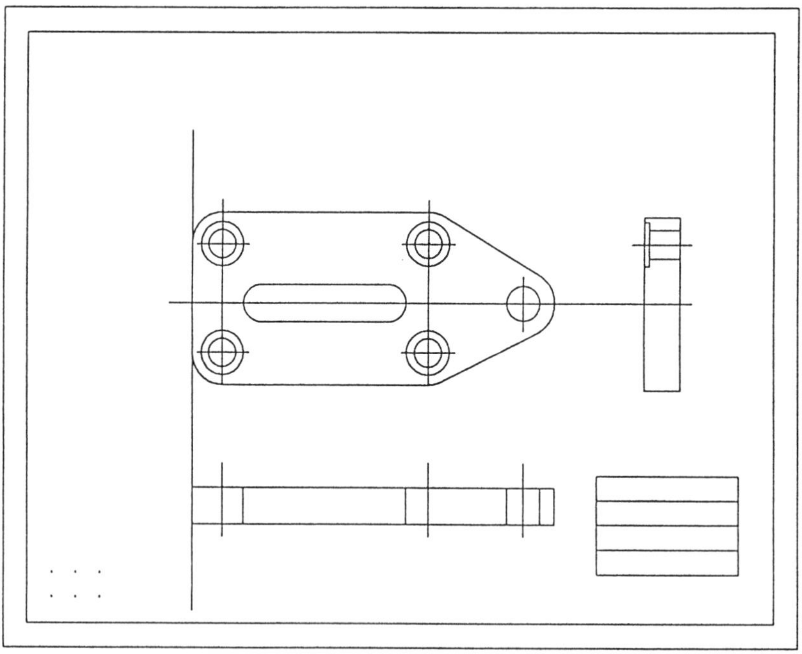

Figure 3.8 Original drawing with missing annotation.

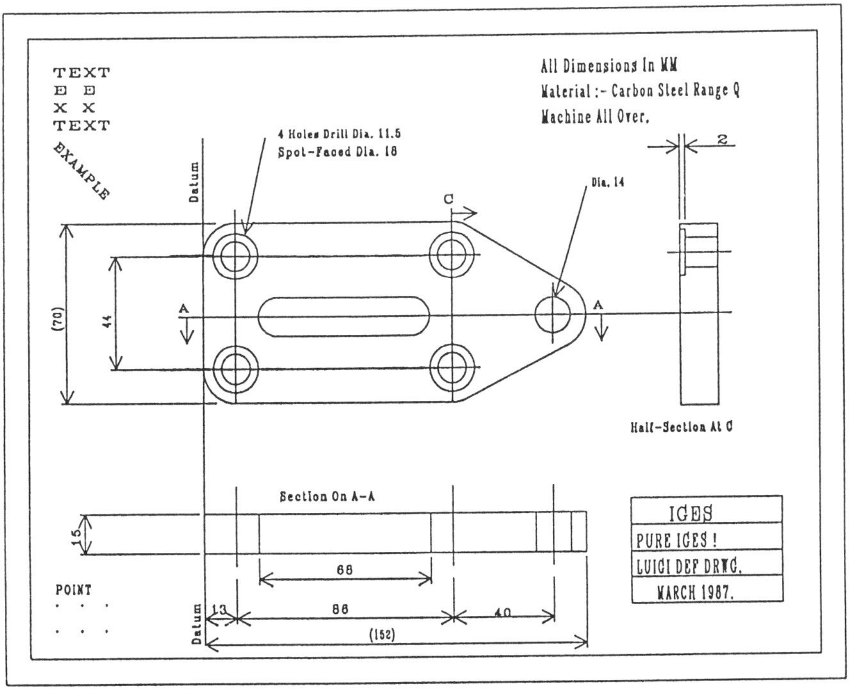

Figure 3.9 Original drawing with incorrect line styles.

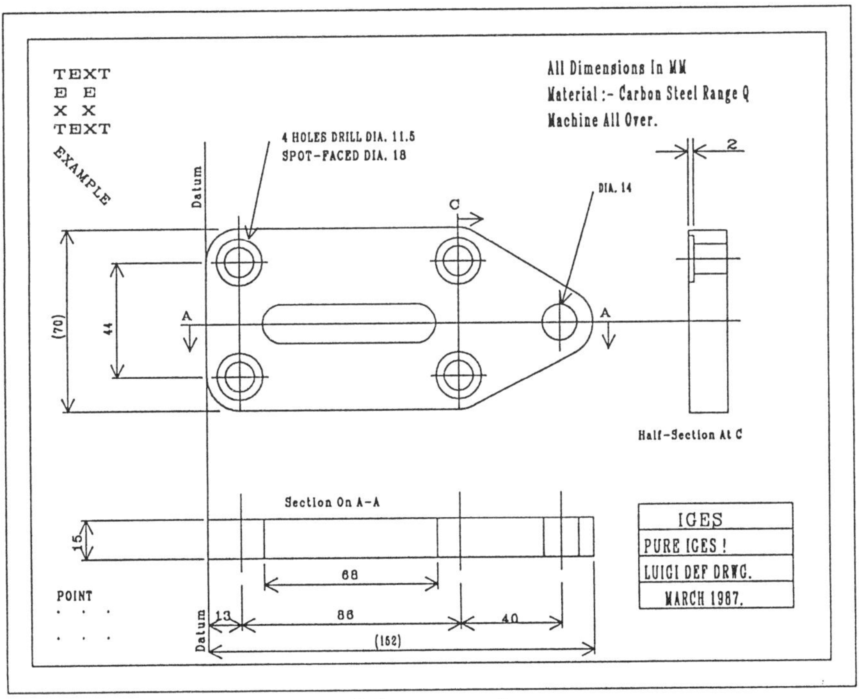

Figure 3.10 Original drawing with incorrect text.

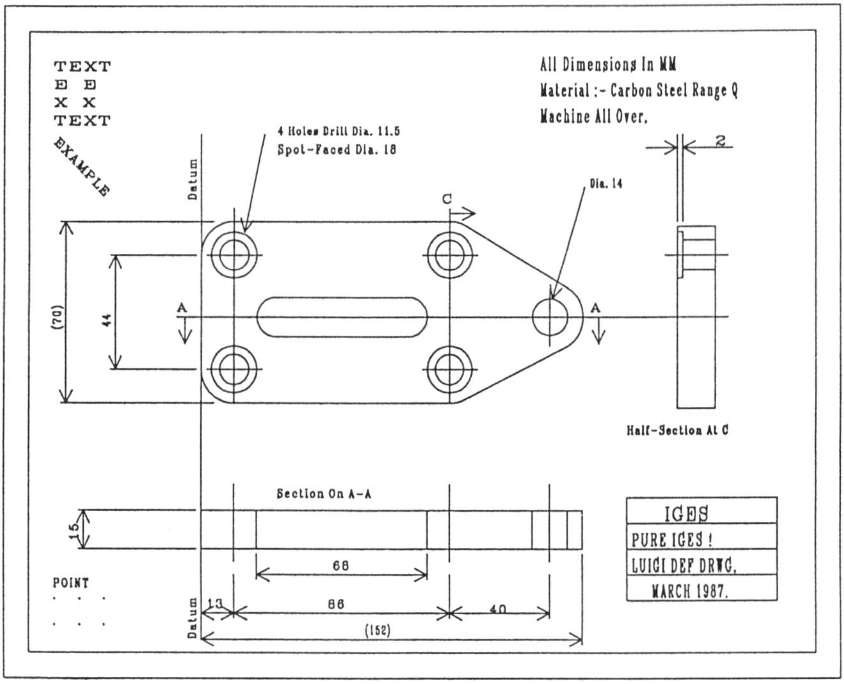

Figure 3.11 Original drawing with unpreserved structure.

transfer of such geometry to an application such as NC machining difficult, if not impossible.

Figure 3.10 has some characters changed from lower to upper case which, for a human interpreter, is not a problem (provided the text is "readable"). However, if all text characters undergo a change to capitals, then a problem might ensue with some of the special (ASCII non-printing) characters used for tolerancing.

The final figure in this series, Figure 3.11 has no visual problem whatsoever: the drawings produced from the files are identical. However, this may conceal a fundamental problem. Whereas the sending system might have grouped the lines and text into particular views, retaining a three-dimensional model and projecting it into the conventional views, the receiving system could have lost this information by converting the projection into two dimensions. This can only be determined by examining the CAD models in the respective systems: the two drawings necessarily do not provide sufficient information.

3.7 Conclusion

In conclusion, the problems it is fair to lay on IGES 1.0 and 2.0 are:

(a) no definition of what constituted an IGES implementation (i.e. how many entities and which set of entities);

(b) poor specification, as far as readability and occurrence of ambiguities are concerned;

(c) large and involved file structure due to "card" format and cross-checking data;

(d) the reproduction of "attribute" information for each individual entity, when it may not be appropriate or be similar across the entire file;

(e) some inadequacies in coverage, even in the draughting and surface domains which it was intended to support.

Many of these arose because IGES was the pioneering work in product data exchange; some of the background has been described [90]. Different views of this topic are available elsewhere [89, 119, 50, 110, 165]. The next chapter deals with the responses to difficulties in using IGES in the early 1980s. The first two of the above problems are addressed in more detail in Part III, because of the impact they had on the development of STEP. It was stated as early as 1982 [90] that "IGES is a baseline from which we can derive immediate and continuing functional benefit while gaining the practical knowledge needed for more advanced efforts".

Chapter 4

The proliferation of product data exchange initiatives

In spite of its faults, IGES did not die. The response was that more people became interested in the IGES committees and so they multiplied and flourished. Work continued on Version 3.0, which is described in Chapter 5. Further, a handbook of Recommended Practices for implementors was published and Request For Change procedures were instituted, as described in Chapter 7.

Other countries demonstrated for the first time an active interest born of a real need for effective engineering drawing and surface data exchange. Several formats which addressed the deficiencies of IGES were proposed at this time. Their place is indicated in the chronology, shown in Figure 0.1, and they are described briefly in this chapter.

4.1 VDA-FS 1.0

VDA-FS 1.0 (Verband der Automobilindustrie Flächen Schnittstelle 1.0) was developed in Germany for the exchange of surface data. It was published on 1 May 1984 [151]. IGES 1.0 was deficient in this domain because its polynomial curve and surface form allowed only cubic (i.e. third-degree or fourth-order) polynomials. Although there is some virtue numerically in using low-order polynomials for curve and surface interpolation, CAD systems often allowed higher-order polynomials to obtain a better fit. So VDA-FS was designed to fill a gap in IGES.

VDA-FS is used to transfer free-form surface data and was defined by a working group of the Association of German Car Manufacturers (VDA). Members of the working group included representatives from the following companies:

Volkswagen	Porsche	BMW
Audi	Daimler-Benz	Technical Höchschule Darmstadt
Bosch	Opel	

VDA-FS 1.0 was deliberately limited to transferring essential elements and keeping the interface simple, thereby allowing transition to national and international standards. It is limited to points with optional related vectors and parametric polynomial curves and patches of any order.

VDA-FS 1.0 is conceptually simple. Units are always millimetres and a right-handed rectangular co-ordinate system is assumed. It has two classes of entity:

(a) *Geometric entities*

point: Cartesian point in three dimensions;

point sequence: enables curves (defined by points) and polygons to be transferred;

point vector sequence: used to transfer points and related vectors for the purposes of numerical control;

curve: used to represent any piecewise polynomial, irrespective of the manner in which it was originally defined (e.g. Bézier, B-spline, Coons [11, 30, 52]);

surface: this information is in the same form as that for the curve, except that information has to be supplied for the patches in both *u*- and *v*-directions.

(b) *Non-geometric entities*

header: the first line in the file (it includes the file name, who created the data and where, and other such identification data);

comment: may be included anywhere in the file;

set: allows geometric data to be combined logically (only a single level is allowed);

trailer: the last record in the file, with the name corresponding to that in the header.

The names and types of the attributes of each of these entities are given in Appendix F.

All parameter data is of the form

```
name "=" command_word [ "/" (number | name) { ","
                        (number | name) } ] .
```

in Wirth Syntax Notation [174] (see Appendix C). Command words represent the entity types; numbers are either reals or integers; and names "point" to other entities.

An example of a VDA-FS file is shown in Figure 4.1. VDA-FS was adopted as German national standard DIN 66301 in July 1986 [152], gained relatively wide acceptance and was implemented quickly by an impressive list of vendors [134]. Its simplicity and the need to exchange

```
VDA001 = HEADER/20                                                      00000100
************************************************************************00000200
VDAFS version 2.0                                                       00000300
----------data about the sending company--------------------------------00000400
sending company: Department of Mechanical Engineering                   00000500
sender's name: Jon Owen and Susan Bloor (iso@uk.ac.leeds.leva)          00000600
sender's telephone number: + 0113 233 2113 (fax 2150)                   00000700
sender's address: The University of Leeds, Leeds, LS2 9JT, UK           00000800
generating system: hand-crafted systems plc                             00000900
creation date: 22nd July 1994                                           00001000
file name: vdafs.file                                                   00001100
----------data about the part-------------------------------------------00001200
project name: Product data exchange book                                00001300
object description number: not available                                00001400
variants: none                                                          00001500
confidentiality: public domain                                          00001600
date effective: immediate                                               00001700
----------data about the receiver---------------------------------------00001800
receiving company name: UCL Press, University College London, London    00001900
receiver: Sheila Knight, Publications                                   00002000
************************************************************************00002100
$$                                                                      00002200
$$ all elements are from VDA-FS version 1.0                             00002300
$$                                                                      00002400
SET01 = BEGINSET                                                        00002500
$$                                                                      00002600
$$ point:                                                               00002700
$$                                                                      00002800
P001 = POINT / 0.0, 0.0, 0.0                                            00002900
P002 = POINT / 0.0, 0.0, 100.0                                          00003000
$$                                                                      00003100
$$ point sequence:                                                      00003200
$$                                                                      00003300
PS001 = PSET / 2,                                                       00003400
               0.0, 0.0, 0.0,                                           00003500
               0.0, 0.0, 100.0                                          00003600
$$                                                                      00003700
$$ point and vector sequence:                                           00003800
$$                                                                      00003900
MDI01 = MDI / 2,                                                        00004000
              0.0, 0.0, 0.0,      1.0, 0.0, 0.0,                        00004100
              0.0, 0.0, 100.0,    0.0, -1.0, 0.0                        00004200
$$                                                                      00004300
$$ curve:                                                               00004400
$$                                                                      00004500
K01 = CURVE / 2, 0.0, 1.0, 2.0,                                         00004600
                5, 1.0,  4.0,   9.0,  0.0   -7.0,                       00004700
                   1.0, 16.0, -33.0, 38.0, -12.5,                       00004800
                   1.0,  8.0,  12.0, -8.0,  -9.0,                       00004900
                3, 7.0, -3.0, 9.0,                                      00005000
                   9.5,  7.0, 0.0,                                      00005100
                   4.0, -4.0, 0.0,                                      00005200
$$                                                                      00005900
$$ end of version 1.0 examples                                          00006000
SET01 = ENDSET                                                          00006050
$$                                                                      00006100
$$ example of a transformation matrix:                                  00006200
$$                                                                      00006300
M01 = TMAT / 0.5, -0.866, 0.0,    0.866, 0.5, 0.0,    0.0, 0.0, 1.0,    00006400
             100.0, 0.0, -10.0                                          00006500
VDA001 = END                                                            00006700
```

Figure 4.1 Example of a VDA-FS file.

such data weighed more heavily than the loss of design intent in VDA-FS transfers.

The five geometric entities may be mapped onto IGES [122] and work is continuing on conversion between different surface representations while maintaining control over error propagation. This basic problem of surface exchange, that of loss of accuracy in converting between polynomials with different bases, between polynomials and procedural representations of surfaces and even between the same base polynomials of different degree, is still exercising many minds. Dannenberg and Nowacki [37] were among the first to report their work in the data exchange context, fired partially by the VDA initiative.

4.2 SET

SET (Standard d'échange et de transfert) was started at the same time as VDA-FS, in 1983, as a response to IGES. Its development was initiated by Aerospatiale, the French aerospace company, whose main complaint about IGES was that it was too large (i.e. the files that were generated were too large simply because of the IGES file format) and that processing took too long as a consequence. It was also believed that IGES did not allow the complete transfer of all the data generated by aerospace companies. The French emphasized more than the early IGES documents a need for a format for data storage and particularly archiving. There are examples of aircraft being maintained fifty years after their initial design. The name "SET" embodies the words "exchange" and "transfer", emphasizing the need for one-way transfer (archiving) as well as two-way exchange.

SET addresses the problem of file size and structure rather than coverage. At that time, it still concentrated on the exchange of engineering drawings and the functionality included geometry, annotation and relationships between entities. Its reduction in file size and complexity was largely due to an in-built hierarchical form of blocks and sub-blocks rather than the logical structure supported by pointers that IGES favoured. The file structure of SET allows sharing of data between

Table 4.1 SET block classes.

Number	Description
0	geometric primitives
1	composite geometric entities
2	graphical representations
3	general blocks
4	drafting applications
5	structural relations
6	connectivity applications (schematics)
80	reserved for users
90	SET management

Table 4.2 SET sub-block classes.

Number	Description
0	geometric definition: result
1	geometric definition: construction
2	graphical representations
3	general sub-block
4	drafting applications
5	structural relations
6	connectivity applications (schematics)
80	reserved for users
90	SET management

Table 4.3 SET dictionary contents.

Number	Description	Number	Description
1	coordinate system	63	graphical approximation of parametric surfaces
2	mode (model or draw)	64	graphical approximation of parametric surfaces
3	type or coordinates	65	graphical approximation of parametric surfaces
4	depth	101	primary character font
5	subordinate block switch	102	character thickness
6	deletion code	103	character height
7	selection code	104	character width
8	level	105	space between characters
9	label	106	character slant angle (italic)
10	label subscript	107	scale factor
11	library name	108	average width
12	dimensional tolerance	109	secondary character font
51	graphical representation	111	text justification
52	blank status	112	text rotation angle
53	line font	113	direction of writing
54	line weight	114	line spacing
55	scale flag	115	mirror flag
56	pen number	201	type of arrowhead or symbol
57	colour code	202	height
61	graphical approximation of curves	203	width
62	graphical approximation of parametric surfaces	204	diameter

records and so reduces the size. This structure was acknowledged to have advantages over that of IGES whose complexity added to the difficulty of the processor implementors.

The form of SET is based on blocks and sub-blocks, with a dictionary. Both blocks and sub-blocks are divided into classes, as shown in Tables 4.1 and 4.2. Each block represents a concept recognized by CAD systems (such as view, point, group) and is defined by several items of information, defined by sub-blocks. A sub-block shown by an asterisk in Table 4.4 is one whose presence may be repeated; those in parentheses are alternatives.

Because the sub-blocks are defined separately, they may be used in the definition of more than one block. For example, the blocks for "circular arc" and "radial dimensioning" (see Table 4.4) both use the sub-block for "parameters of a circular arc" (see Table 4.5). The names and types of the attributes of each of the sub-blocks are given in Appendix G.

Whereas in IGES, attributes were given for each entity in its directory entry, in SET such attributes may be defaulted, or set for the whole of a particular assembly or sub-assembly. These attributes are given in the SET dictionary, as shown in Table 4.3.

Table 4.4 SET block types and sub-block usage.

Number	Description	Sub-block usage
Geometric primitives		
1	point	1
2	line	2
3	linear string	3
4	vector	4
10	circular arc	10
11	elliptical arc	11 (14)
12	parabolic arc	12 (14)
13	hyperbolic arc	13 (14)
20	parametric curve	20 21* 9911
30	plane	30 (31)
36	surface of revolution	130
37	tabulated cylinder	131
38	ruled surface	132
40	parametric surface	40 41* 9911
Composite geometric entities		
120	composite curve	120
Graphical representations		
215	view definition block	401 310 216 (251)
220	drawing definition block	220 221* (251)
250	line font	253* 254*
251	definition of a graphical symbol	257
252	character font definition	255*
General blocks		
301	co-ordinate system definition block	301
310	level list block	312
Drafting applications		
400	TNODE	1 400 510 510 409
401	text	401* 510 (409)
402	note	401* 409
412	linear dimensioning	408* 406* 402* 409
413	angular dimensioning	408* 407* 402* 409
414	radial dimensioning	406 402* 409 10
415	diameter dimensioning	406 402* 409
416	label	401 402 406* 409
421	centre lines	410
422	centring axis	410
430	cross-hatching	430*
Structural relations		
500	calling block (internal or external)	500 313 or 314 or 315 or 316 or 317 (510) (520)
510	group	510*
520	attribute	520 or 521
Connectivity applications (schematics)		
601	connection node (CNODE)	519 510 510* 515* 521*
602	connective link (CLINE)	602 510 510* 515* 519 521*
605	connective component	519 610 510 510* 515 521*
Reserved for users		
80xx	user-defined	user-defined
SET management		
9900	start of SET	9900 9901 9902*
9901	start of assembly	9905 9920 9930* 9900 9901 9902* 520*
9902	start of sub-assembly	9905 520* 9900 9901 9902*
9997	end of sub-assembly	none
9998	end of assembly	none
9999	end of SET	none

Table 4.5 SET sub-blocks

Number	Description
Geometric definition: result	
1	point co-ordinates
2	straight line segment
3	polygonal line
4	vector components
10	parameters of a circular arc
11	parameters of an elliptical arc
12	parameters of a parabolic arc
13	parameters of a hyperbolic arc
14	coefficients of a conic
20	start of a parametric polynomial curve
21	definition of a polynomial curve segment
30	cartesian definition of a plane
31	parametric definition of a plane
40	start of a parametric polynomial surface
41	definition of a polynomial patch
Geometric definition: construction	
120	definition of a composite curve
130	parametric definition of a surface of revolution
131	parametric definition of a tabulated cylinder
132	parametric definition of a ruled surface
Graphical representations	
216	projection method
220	definition of drawing space
221	display of view on a drawing
251	graphical representation of a list of blocks
252	graphical representation of a block in a list of views
253	linear plot (pen)
254	linear positioning of symbols
255	association of ASCII code and character definition
257	graphical representation of a symbol
258	graphical representation of a wireframe entity (variable line font)
General sub-block	
301	reference system's transformation matrix
310	windowing
312	list of levels
313	definition of a 3-D transformation
314	general matrix of a 3-D transformation
315	definition of a 2-D transformation
316	general matrix of 2-D transformation
317	translation
Drafting applications	
400	TNODE parameters
401	text description
402	dimension text
406	linear dimension line
407	angular dimension line
408	witness line
409	box (framework)
410	axis lines
430	cross-hatching
Structural relations	
500	call definition
510	associativity defined by SET
515	associativity towards representation
519	instance name and technical reference
520	attributes defined by SET
521	attributes defined by user
Connectivity applications (schematics)	
602	logical link between two CNODES
610	list of component terminals
Reserved for users	
80xx	user-defined
SET management	
9900	file identification
9901	date
9902	comments
9905	model space definition
9911	last sub-block in a series of sub-blocks
9920	initialization of dictionary parameters
9930	header block descriptor

```
@9900
  #9900, 'Z68-300.85-08', 'interface version 1.0',
         'Hand-crafted systems plc', 'software version 1.0',
         'The University of Leeds', 'Faculty of Engineering',
         'Department of Mechanical Engineering', 'Jon Owen'
  #9901, 1985, 08, 01
  #9902, 'Example SET file'
@9901, 1, 'pyramid'
  #9905, 2, 1.0, 1, 1, 2, -10.0, -10.0, 10.0, 10.0
  #9920,
  #9930,
  #9901, 1992, 08, 19, 09, 00, 00
@2, #2  -10.0, -10.0, -10.0, +10.0, -10.0, -10.0
@2, #2  +10.0, -10.0, -10.0, +10.0, +10.0, -10.0
@2, #2  +10.0, +10.0, -10.0, -10.0, +10.0, -10.0
@2, #2  -10.0, +10.0, -10.0, -10.0, -10.0, -10.0
@2, #2  -10.0, -10.0, -10.0,   0.0,   0.0, +10.0
@2, #2  -10.0, +10.0, -10.0,   0.0,   0.0, +10.0
@2, #2  +10.0, -10.0, -10.0,   0.0,   0.0, +10.0
@2, #2  +10.0, +10.0, -10.0,   0.0,   0.0, +10.0
#9998
#9999
```

Figure 4.2 Example of a SET file.

An example SET file, shown in Figure 4.2, defines a pyramid and uses the values from the default dictionary. It has been laid out so that the structure of the blocks (@) and sub-blocks (#) is clear. The blocks of types 9900 and 9901 form the header; those of type 2 define several lines; and those of types 9998 and 9999 form the trailer of the file. The number after each block type is its serial number, except for that at the end of the file, which is the total number of blocks in the file. A more complete example, using the 1989 version of SET, is available from Association GOSET [70].

The generation methods for sub-blocks 20 and 40 are as follows:

0. unspecified;
1. Bézier;
2. uniform B-spline;
3. rational B-spline;
4. spline;
5. Wilson–Fowler.

Thus, SET included the Bézier and rational B-spline forms for curves and surfaces which IGES 1.0 had omitted; this was rectified in IGES 2.0.

SET was published in 1984 [136] and was used to transfer data between companies in the European Airbus project. It became French standard z68-300 in August 1985 [137]. It was purported to process three times faster than a comparable IGES file and be ten times shorter when tested on a layout drawing of an Airbus.

An important advantage that SET had over IGES was the strong paternal interest which Aerospatiale retained in its development. This included promoting SET and also guiding development. The provision of X-SET, which is software to read and write a SET file, pre-empted the need for every CAD vendor to write similar code and predated the CADEX approach to processor generation by several years.

4.3 XBF

When McAuto's work on solid modelling was rejected by CAM-I, the ideas were incorporated in XBF (Experimental Boundary File) [173] which was based on the IGES file format. XBF is a format for the transmission of solid modelling data. It defines new geometric and topological entities, but in an IGES-compatible manner (i.e. it uses the same syntax for new entities and new forms of existing IGES entities were defined where appropriate). At the same time as CAM-I contracted the project to McAuto they also chose Shape Data, a British CAD firm with solid modelling experience, to implement a procedural interface which was known as the Application Interface System, AIS [173]. This was an interface between functional applications and boundary representation solid modellers. It consisted of some 150 callable subroutines satisfying geometric modelling. XBF was available in 1982 and was demonstrated with the exchange of boundary representation solid models along with the procedural interface.

4.4 PDDI

PDDI (Product Definition Data Interface) [113] was another project initiated by the United States Air Force ICAM project. It was led by McAir (McDonnell Douglas Aircraft Corporation) and strove to understand the issues in generating complete computer-based product definition data for use between design and manufacture. It includes the scope addressed by IGES but also considered manufacturing information such as features relevant to machining, turning and sheet-metal handling and tolerancing, geometric tolerancing and datum positions. Four representative types of part, typical of those from the airframe industry, were used as case studies.

The project was divided into two tasks. The first assessed the general usability of IGES in CAD graphics, for which interim results were presented to the IGES Committee at the meeting held in October 1983. They reported that "conic translation is a widespread source of problems", due to divide-by-zero errors when a conic was nearly linear, scaling problems with ellipses, and single- and double-precision representation causing changes in the form and accuracy of the conic. These problems are addressed further elsewhere [164].

The second task attempted to understand the manufacturing data requirements from engineering, to establish system needs and to design, implement and demonstrate a PDDI prototype based on process planning

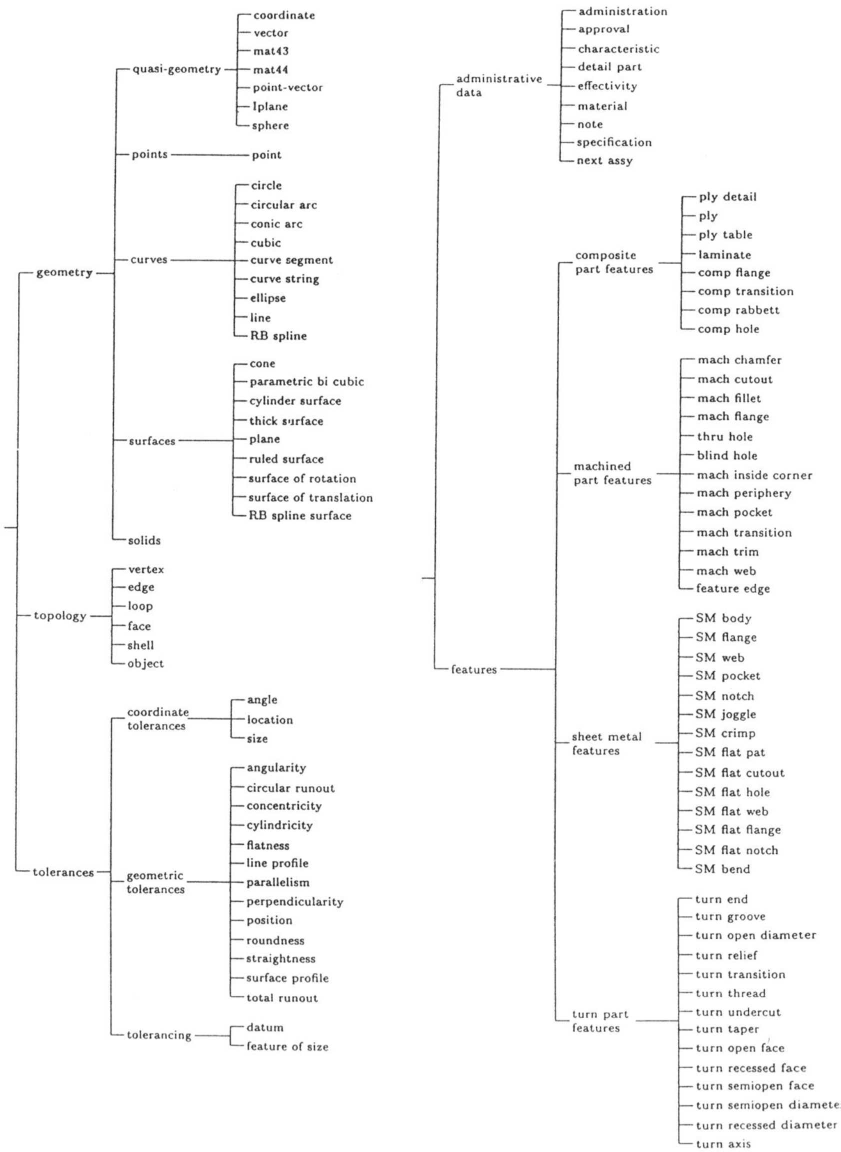

Figure 4.3 PDDI entities.

and NC manufacture. They identified the important, but not even yet properly formalized, data elements as geometry including boundary representations, dimensional and geometric tolerances, part features and part control. The entities and their classification are shown in Figure 4.3.

The PDDI exchange format file has seven sections, of which only the first two and last one are mandatory:

(a) metadata;
(b) format;
(c) explanation;
(d) characteristics;
(e) statistics;
(f) relationships;
(g) data.

Records denoted as "explanation" may be placed anywhere in the file, as comments. The characteristics section includes information on number representation and units, similar to some of the IGES global fields. The statistics give the totals of each entity type and information concerning external references. The final two sections are used to define the product definition data itself, and the relationships that these entities have (such as associations). Each line is identified by a letter at its start, indicating its type.

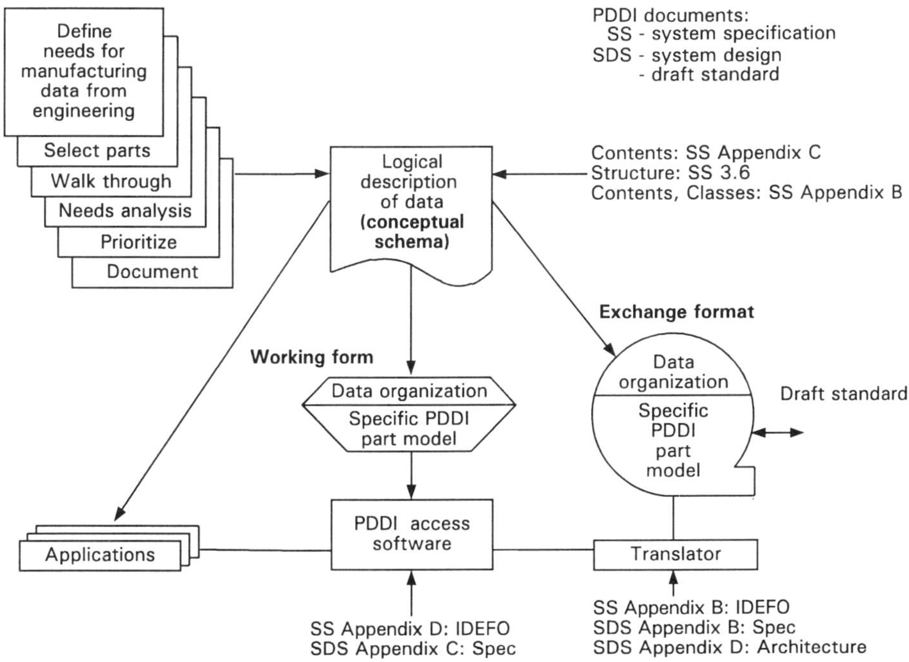

Figure 4.4 PDDI system architecture.

The most important feature of the PDDI exchange format were the first two sections. The metadata section (i.e. data about data) defines the syntax, domain and validity of the entities in all of the following sections. If a standard set of entities were being used, a predefined metadata section could be named, but this facility also allowed entities to be added, made unavailable, or modified. The section was derived from the conceptual schema. The format section contained the tape format of all sections. In addition, a different file structure (such as IGES) could be defined.

A conceptual schema was defined in a data definition language, which was both human-readable and computer-processable. The working form (a memory-resident form of the data) and the exchange format (a physical file format) were both derived from it. Model access software was written to provide an interface between the working form and applications, and was used in the writing of translators between the working form and the exchange format. The PDDI architecture is shown in Figure 4.4, and is reminiscent of the application interface specification (AIS) which CAM-I proposed and demonstrated.

It is understood that PDDI was demonstrated but we know of no documented results. PDDI had a major influence on the description language and physical file proposals for STEP. It also contributed to the understanding of the full extent of product definition data which, through the IGES research project PDES described in Part IV, has also influenced STEP.

4.5 Conclusion

The publication of IGES 1.0 spawned many national efforts to define specifications for product data exchange, although much of the early work was undertaken in isolation.[2] By the end of 1983, standards makers, CAx system vendors and users were putting considerable effort into making data exchange a reliable process. Part III explains the difficulties in solving the problems, which were by now well identified. Chapter 5 concentrates upon the developments made subsequently to the established standards and specifications for product data. Chapter 6 provides an overview of other standards and specifications, relevant for computer-aided engineering, which also contributed to the field of product data exchange.

Chapter 5

Further developments

The STEP project was initiated in 1984: this is described in Part IV. Consequently, work on the national initiatives described in this chapter continued in parallel with its development. This benefited both STEP and the national initiatives, in that expertise was available for both. The national initiatives could include facilities expected in STEP, and be used to test ideas, whereas STEP would learn from the mistakes made with the early specifications.

5.1 IGES

As promised, work continued on improving IGES following Version 2.0 which was accepted and used by the product data exchange community. However, it was not submitted for standardization. IGES 2.0 was published in 1983, just over two years after the first appearance of IGES 1.0. The publication of IGES 3.0 was anticipated in 1985 but did not appear until March 1986 [78].

The additional entities and forms in IGES 3.0 are summarized in Table 5.1. Broadly, the changes were as follows.

Domain
IGES 3.0 included more finite-element and electrical extensions. It also made provision for geometric elements of use to manufacture, such as offset surfaces and curves on surfaces.

External references
It also allowed external references to be made; this was necessary to allow standard symbol and component library files to be used. This was

Table 5.1 New entities and forms in IGES 3.0.

Entity	Form	Description	Entity	Form	Description
124		transformation matrix	314		colour definition
	1	orthonormal rotate – reflect/translate	320		network subfigure definition
130		offset curve	402		associativity instance
132		connect point		2	external logical reference file index
138		nodal displacement and rotation		12	external reference file index
140		offset surface		13	dimensioned geometry
142		curve on a parametric surface		14	ordered group with backpointers
144		trimmed parametric surface		15	ordered group without backpointers
212		general note		16	planar group
	1	dual stack		18	flow
	2	imbedded font change	406		property
	3	superscript		11	tabular data
	4	subscript		12	external reference file list
	5	super-/subscript		13	nominal size
	6	multiple stack / left justified		14	flow line specification
	7	multiple stack / centre justified		15	name
	8	multiple stack / right justified		16	drawing size
	100	simple fraction		17	drawing units
	101	dual stack fraction	416		external reference
	102	imbedded font change / double fraction		0	one definition from library file
	105	super-/subscript fraction		1	full referenced library file
214		leader arrow		2	external logical references
	11	open triangle	418		nodal load and constraint
228		general symbol	420		network subfigure instance
	0	general symbol	5001 . . . 9999		implementor-defined macro instance
230		sectioned area	10000 . . . 99999		implementor-defined macro instance
	0	standard cross-hatching			
312		text display template			
	0	absolute co-ordinates			
	1	incremental co-ordinates			

a much criticised omission from earlier versions of IGES which required that all information relevant to an IGES transfer (e.g. a drawing) had to be in one physical file. This meant that one had to define and instantiate symbols within the file which contained the drawing and interpret connection manually with any others used. The long-promised extensions from the AEC committee included a facility for tabular data, treatment of flow lines in a similar way for electricity and fluids, and connectivity relationships (as in piping and instrumentation diagrams and electrical schematics).

File compression

There was added potential in IGES 3.0 for file-store reduction using a compressed ASCII format which changed the form of both the directory entry and the parameter data sections. Utilities written in Fortran for converting between the ASCII and compressed ASCII forms were included in the document.

Documentation

The IGES 3.0 document generally had better explanations than earlier IGES documents, contained improvements to ensure numerical stability in some of the geometry entries, and had good examples in the appendices indicating the entities to be used for particular applications. This last is of utmost importance if implementors are to interpret any specification consistently and thus give users a chance of exchanging information. As will be seen, this type of explanation is an integral, formalized part of STEP.

The next version of IGES, Version 4.0 [79], was long anticipated because it was to contain provision for the exchange of solid models. Since the very early work of CAM-I culminating in the IGES-compatible XBF format, progress had continued under the aegis of the IGES Advanced

Table 5.2 New entities and forms in IGES 4.0.

Entity	Form	Description	Entity	Form	Description
0		null	222		radius dimension
108		plane		1	multiple leader format
	0	unbounded	228		general symbol
146		FEM nodal results		1	datum feature
148		FEM element results		2	datum target
150		block		3	feature control frame
152		right angular wedge	322		attribute table definition
154		right circular cylinder			
156		right circular cone frustrum		0	one-to-many
				1	one-to-one
158		sphere		2	one-to-one with text display template
160		torus			
162		solid of revolution	406		property
	0	closed to axis		18	inter-character spacing
	1	closed to itself	422		attribute table instance
164		solid of linear extrusion		0	single row
168		ellipsoid		1	multiple row
180		boolean tree	430		solid instance
184		solid assembly			

Modelling Committee in their Experimental Solids Programme (ESP). The results of this programme fed into IGES 4.0, IGES 5.0 and also the STEP solids provision.

The additional entities and forms in IGES 4.0 are shown in Table 5.2. The major changes were in the following areas.

Solids

In the event, IGES 4.0 was disappointing in this respect because provision for boundary representation solids was delayed until a future version. IGES 4.0 contained only entities to support the exchange of constructive solid geometry; that is, it contained solid primitive entities and the means of defining their combination.

Other functionality

IGES 4.0 extended once more the provision for finite-element data by specifically representing results. It enhanced the provision for attribute data and graphic representation for architecture and engineering construction (AEC) applications.

Documentation

One innovation was the inclusion of "grey" pages; these contain entities which are not considered to be in the full specification until at least three vendors have implemented them and transferred them success-

Table 5.3 New entities and forms in IGES 5.0.

Entity	Form	Description
141		boundary
143		bounded surface
182		selected component
204		curve dimension
213		new general note
214		leader (arrow)
	12	dimension origin
216		linear dimension
	1	diameter form
	2	radius form
218		ordinate dimension
	1	witness line and leader
228		general symbol
5001..9999		implementor-defined
230		sectioned area
	1	inverted cross-hatching
316		units data
402		associativity instance
	19	segmented views visible
404		drawing
	1	translate, scale and rotate
406		property
	19	line font
	20	highlight
	21	pick
	22	uniform rectangular grid
	23	associativity group type
	24	level to PWB layer map
	25	PWB artwork stackup
	26	PWB drilled hole
410		view
	1	perspective
416		external reference
	3	native system subfigure

fully. Another was the availability of the document itself in electronic form as a series of LaTeX [88] files.

IGES continues to be improved and facilities which took longer to agree than expected are still being added. IGES is a useful proving ground for ideas which are being incorporated into STEP. The new entities and forms in IGES 5.0 (which appeared in Autumn 1990 [80]) are presented in Table 5.3. The major inclusions were as follows.

Conformance
The first conformance clauses were included so that implementers knew what their preprocessors must produce and postprocessors must read to fulfil the requirements of the specification.

Application protocols
Support for application protocols was included (see Ch. 9) and in Volume 2 of IGES the first technically complete and approved application protocol (for piping applications).

Documentation
The presentation of the document was again improved. This was achieved by including entities in numerical order, a directory entry table with each entity showing the permitted values for the fields, and clarification of default values.

Units
In earlier versions of IGES, only a single global unit could be specified for geometric entities (using global fields 14 and 15). A special case also existed for specifying tabular data, in which case the SI defaults were applied. However, up to this version of IGES, the user had only global control over units. In IGES 5.0, improvements were made in the use of units, so that a particular unit could be attached to a given entity: this is particularly useful in AEC, in which lengths of roads or pipes may be several hundred metres in one dimension and several orders of magnitude smaller in the others.

File format
The binary form, introduced in IGES 2.0, was deprecated in IGES 5.0. In practice, a reduction in hardware costs and provision of operating-system compression utilities had helped to alleviate this problem; the IGES compressed ASCII form was now available too.

IGES 5.1 was published in 1991 [81] and included the long-awaited provision for boundary representation of solid models. The new entities and

Table 5.4 New entities and forms in IGES 5.1.

Entity	Form	Description	Entity	Form	Description
123		direction	198		toroidal surface
186		manifold solid B-rep object		0	unparameterized surface
190		plane surface		1	parameterized surface
	0	unparameterized surface	402		associativity instance
	1	parameterized surface		20	piping flow
192		right circular cylindrical surface		21	dimensioned geometry
	0	unparameterized surface	406		property
	1	parameterized surface		27	generic data
194		right circular conical surface		28	dimension units
	0	unparameterized surface		29	dimension tolerance
	1	parameterized surface		30	dimension display data
196		spherical surface		31	basic dimension
	0	unparameterized surface	416		external reference
	1	parameterized surface		4	native library subfigure
			502		vertex
			504		edge
			508		loop
			510		face
			514		shell

forms are shown in Table 5.4; the other major changes were that provision was made in the attribute tables for piping information and dimension attributes were categorized as manual, generative or automatic.

5.2 VDA-FS

Version 2.0 of VDA-FS first appeared in 1986 but was superseded by the version published on 1 January 1987 [153]. It allowed the representation of a circle (which cannot be represented exactly by simple polynomials), curve on surface, face, topology, group, matrix and transformation list. Their attribute names and types are described in Appendix F, and an example of their use is presented in Figure. 4.1.

The original developers, listed in Section 4.1, were supplemented by EDS, Hella, Ford and Technische Universität Berlin. Following VDA-FS 2.0, the German group responsible for the format agreed that they would contribute all their efforts towards defining the new standard STEP and so work on VDA-FS was halted.

VDA-FS 2.0 has an impressive list of implementors including most of the vendors of CAD systems that include surface facilities and who are based in the United States as well as European vendors. The format is in use among car manufacturers and their suppliers. Its success demonstrates the need for electronic data exchange and the value of simplicity over flexibility which carries with it complexity. Ultimately, the need to transfer data overrides that of complete information content. However, it must be remembered that the representation of surfaces is still mainly a means of providing large amounts of data for downstream activities such as visualization and manufacture. Surface design, either from a functional specification or where the criteria are aesthetic, is still not done optimally on the computer. This is because the parameters are those of the polynomials representing shape rather than those directly aligned to the function of the component.

5.3 SET

Following standardization of SET 1.0 in 1985 and active support and promotion by its development team in Aerospatiale aided by demonstrable deficiencies in IGES, the use of SET in France increased. Development of SET 2.0 was controlled by a group consisting of defence interests and the automobile manufacturers as well as Aerospatiale. This group became GOSET, an organization financed by subscription from member companies. This is a well managed development team containing many of the original Aerospatiale SET team.

Work on extending SET closely parallels the extensions to IGES in the need to provide for the information transfer needs of more complete geometry, different applications and more relationships. Provision for solid modelling, more surface forms and finite-element data were added, with a new national standard in France published in 1989 [138].

SET version 2.0 (revision B0) was published in January 1991. It contained additional provision for solid models, schematics and scientific data.

5.4 Conclusion

Development of the existing specifications continued, with improvements to documentation and provision for application areas other than engineering drawings and surfaces. IGES included provision for finite elements, electrical applications, AEC, geometry for downstream manufacturing, solid modelling, and the much-needed external reference facility to allow libraries to be used. SET included similar facilities, as well as provision for schematics and scientific data. VDA-FS was also updated, but kept its focus very much on the minimum structures need for use in the automotive sector. Parallel work continued in the electrical sector, particularly in EDIF, and in related standards, mainly those in graphics and networking.

Chapter 6

Related initiatives

This chapter describes very briefly some of the other initiatives related to computer-aided engineering systems which have resulted in the publication of standards and specifications. They are noted for several reasons. First, the methods by which some were developed and tested were used by later versions of product data exchange specifications, particularly STEP. Secondly, some provided an infrastructure in which the product data exchange specifications would be used. Thirdly, and perhaps most importantly, some confusion arose in the minds of users concerning what standards were actually needed in an engineering enterprise. This chapter attempts to dispel that confusion.

6.1 Graphics standards

One of the difficulties encountered in the early promotion of IGES was the confusion people had about its relationship to other graphics standards. This was because it was named the "Initial Graphics Exchange Specification", and because engineering use of computers blossomed with the availability of computer graphics.

The relationship between graphics standards and CADCAM data-exchange standards is shown in Figure 6.1 which was produced for a contribution on data exchange standards in a special issue of the *Computer-Aided Design* journal [110].

Graphics standards themselves are of two forms:

(a) procedural definitions, which are designed to make programs which wish to use graphics devices independent of the hardware (graphics device) environment; and
(b) data formats, which may be used to transfer graphics data between graphics systems.

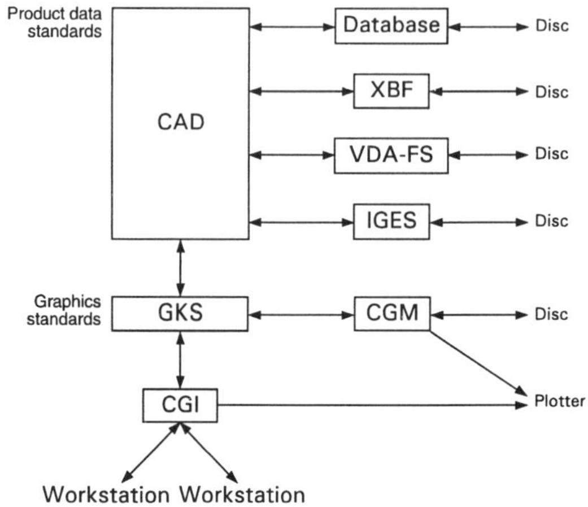

Figure 6.1 The relationship between graphics and product data standards.

6.1.1 Graphics systems

We first deal with the graphics systems for program independence. The forerunners of the standards which are now available were used in the 1970s and their development has reflected graphics hardware developments. There are many fuller descriptions, both *Computer-Aided Design* and *Computer Graphics and Applications* devote issues to graphics standard updates [18, 140, 100]. GHOST, from the Atomic Weapons Research Establishment (AWRE) at Harwell in the United Kingdom, was used for drawing graphs for the scientific and engineering community in UK universities from the mid-1960s until the 1980s. As with today's standards, the user just had to know how to use a set of procedure calls and have some understanding of scaling. Proper orientation and sizing of text was the biggest problem. The output medium was a plotter.

GINO (Graphical Input and Output) came from the CAD Centre also in the UK. It reflected the needs of those in the engineering community who wished to represent artefacts by the transformations and projections that are used in engineering drawings or to use transformations of three-dimensional data that are only easily done by computer (e.g. perspective). The ideas of windows and viewing transformations were included. As well as obtaining hard-copy output, one could operate from graphics terminals, providing input and generating output to the screen. Neither GINO nor GHOST was submitted to the rigours of standardization but their use was widespread in the UK. The community was, however, beginning to realize the cost of transporting programs between computing systems and changing graphics libraries. The graphics community met in a series of meetings that culminated in GKS and also spawned

Core. The history of these meetings is entertainingly documented by Hopgood et al. [69].

Core [63, 64, 51] was an American proposal for an application-independent graphics system for three dimensions which appeared to be gaining some support from vendors of workstations in the mid-1980s. It has now been overtaken by GKS-3D and PHIGS, which are introduced below.

6.1.2 The Graphical Kernel System (GKS)

GKS is an application-independent computer graphics system which provides a functional specification for two-dimensional graphics. It originated in Germany, but was worked on subsequently by the international community. It was standardized in July 1985 [56].

It includes facilities for

(a) device control;
(b) displaying pictures using output primitives (polyline, polymarker, text, fill area, cell array and generalized drawing primitive);
(c) defining attributes which determine the appearance of the output;
(d) graphical input using a graphics device;
(e) transformations;
(f) picture segmentation;
(g) error handling;
(h) archiving.

The GKS document defines the graphical functionality in a computer language-independent form. Thus, each programming language must have a consistent mapping of the GKS functions. This is called a language binding and allows the facilities of a language to be used fully: those languages that allow long identifiers and a range of data structures may allow such full usage. Bindings have themselves become standards [57].

It is not our place to describe GKS fully. Its interest here is threefold. First, the confusion with IGES needs dispelling. As shown in Figure 6.1, GKS is used to perform graphical input and output, whereas IGES is used for capturing product data. One manifestation of such data is on an engineering drawing, which may be produced using lines and text with a graphics system. However, it contains a conceptually higher level of information: an entity such as a linear dimension would be decomposed by a graphics system into its atomic lines and text, and all structure of a model would also be lost. Secondly, GKS was defined using a finite-state machine approach. It was one of the first standards to realize the necessity of moving away from reliance on natural language to define a standard from which computer software would be generated. Lastly, the designers realized that all implementors would not be considering environments with the same functionality, so they introduced the idea of levels. These levels were well defined and are shown in Table 6.1. Each

Table 6.1 GKS levels.

Output level	Input level a	b	c
0	No input, minimal control, only predefined bundles, multiple normalization transformation facilities but minimum settable required is 1, and all output functions; metafile workstation optional	REQUEST input, mode setting and initialize functions for logical input devices, no PICK, and set viewport input priority	SAMPLE and EVENT input; no PICK
1	Full output including full bundle concept, multiple workstation concept, basic segmentation (everything except workstation-independent segment storage); metafile workstation required	REQUEST PICK, mode setting and initialize for pick	SAMPLE and EVENT input for PICK
2	Workstation-independent segment storage		

function in the specification includes a designation of the level to which it belongs. Thus, for a particular level, there is a list of functions that need to be supported. This is an attempt to control the problem that IGES has faced, of undisciplined implementations of those parts of the standard that were convenient to the implementor.

6.1.3 GKS-3D and PHIGS

GKS-3D and PHIGS are international standards which are based on the impetus provided by GKS and on the concepts and methodologies of GKS and Core. GKS-3D [58] adds output primitives in three dimensions and mechanisms for controlling the viewing projections, for input in three dimensions, and for the appearance of primitives to GKS itself. However, it does not change the existing functions for two dimensions. It can be regarded as the minimal update sufficient for three-dimensional functionality. PHIGS [114, 116] has been designed to support the very powerful graphics workstations used for highly interactive applications in which the screen must be updated rapidly – perhaps even dynamically – in response to user control. The eventual outcome of these two initiatives in the light of the increasing availability and use of advanced graphics workstations and the vendor control of the graphics system used at the lower end of the market on PCs is difficult to foresee.

Language bindings have also been defined for GKS-3D [59] and PHIGS [115].

6.1.4 Computer graphics interface (CGI)

CGI [26] was conceived during the development of GKS. It was originally called the virtual device interface and it specifies the interface to a hardware device. This enables device drivers to be written by the graphics hardware vendors more easily. Any such standard has to be sufficiently flexible to address the wide range of intelligence that can appear in a device in firmware and to take advantage of it. Many workstations now do transformations (rotation, zooming) in hardware that has previously been included in a graphics software system.

6.1.5 Graphics metafiles

The issues catered for above are of interest to the graphics programmer. The user of such programs wishes to input data, possibly received from elsewhere in machine-readable form, and send output (a picture) to someone else electronically. The engineering user wants to send engineering drawings; other users wish to send pictures which may only require reproduction on the receiver's output device. When IGES became available an alternative was the Calcomp graphics format which was only able to represent the graphic elements that an incremental plotter could interpret (i.e. to quote Wilson [166] "the detail, but none of the abstraction required").

The computer graphics metafile (CGM) [27] has emerged as the format for the exchange of graphics elements. It differs from the GKS metafile: the latter provides a trailfile of all graphical transactions, whereas the CGM provides a snapshot of the graphical output at a particular time.

6.2 Electronics

The electronics industry has been a trail-blazer in the use of computer-aided design, testing and manufacture. It is not surprising that this sector has felt the need to exchange information. IGES, with its concentration on mechanical CAD did not address its needs. EDIF and VHDL are interesting particularly with respect to identifying their overlap with the IGES schematic representation and their influence on STEP. EDIF is emphasized because of its parallel evolution with IGES and STEP, and its increasing use in the electronics industry.

6.2.1 IPC

The Institute for Interconnecting and Packaging Electronic Circuits (IPC) produced the initial release of D-350 in 1972. It was first used in a contractual requirement by the US Department of Defense a year later, was updated in 1975, and was published as an American standard in 1977 [7]. This allowed the exchange of printed circuit and wiring board data for design, draughting, tooling and manufacturing.

6.2.2 EDIF

In late 1983, the VLSI design and manufacture industry began developments which led to EDIF (Electronic Design Interchange Format) [47]. As its name suggests, it is used to exchange information about a design.

Work on EDIF was carried out initially by a very small group of people who drew heavily on existing proprietary industry formats for the kernel of their proposals. In this sense, EDIF was – and is – a truly industry-driven standard. There were various releases of EDIF which were reviewed by interested parties, until EDIF Version 2 0 0 was released in 1987 and became an accepted EIA (Electronic Industries Association) and ANSI standard.

The focus of EDIF is electronic *design*, and hence Version 2 0 0 provides a wide range of capabilities. These include provision for netlist (handling hierarchy, connectivity, complex objects, bussed ports), schematics, logic modelling (for functional representation of cells), mask layout (for integrated circuit representation) and a form of printed circuit board layout. Soon after the release of EDIF Version 2 0 0, many vendors came forward claiming EDIF-compatible interfaces. Unfortunately, their claims were rather over-ambitious. In practice, the early implementations were naïve and simplistic, and failed to conform to the standard. Hence, EDIF was regarded as incapable of representing the required information. After considerable customer pressure to do better, the second generation of EDIF interfaces produced by vendors was greatly improved, and EDIF is now used widely for the interchange of electronic design data. In reality, concerns over proprietary information limited the use of mask layout information. The PCB layout information was naïvely represented and few vendors saw the need to represent functional models in EDIF with the advent of VHDL and Verilog. So, EDIF filled the industrial need of providing the means of moving connectivity information (netlist) and schematic information between systems and also between systems and vendors. At a recent Design Automation Electronics Conference, there was a very long list of companies who were identified as EDIF users and whose product developments rely on the use of EDIF somewhere in the design and manufacture cycle.

The eventual success of EDIF has led to a very considerable body of users, and to wide industrial use which itself has identified shortcomings in the format as represented by EDIF Version 2 0 0. Therefore, since 1991, the EDIF organization has concentrated on the enhancement of EDIF to meet new requirements and to correct anomalies identified in

Version 2 0 0. In parallel, there have been very active developments in three new areas: particularly PCB Layout, Test and (more recently) a form focused on field-programmable gate array (FPGA) representation (called LPM – library of programmable modules).

The new release of EDIF was published as an EIA standard in December 1993 [48]. It has also been accepted by ANSI. Version 3 0 0 concentrates on connectivity and schematics, which are the two areas in which EDIF is used most widely in industry. In addition, the format has been enhanced to handle parameterized information, including FOR frames and conditional frames.

One of the biggest developments has been the pioneering work on the semi-formal specification of the requirements for the new standard, ahead of the development of a representative syntax. The EDIF organization accepted the technique of information modelling in 1990 as the way forward for the specification of requirements. As a result, the new standard is published both as a formal information model written in EXPRESS and a reference manual and syntax that realize that model. Hence, developers and users have a well documented standard with clear semantics. The idea of using information modelling in this way has been taken on by other workers in the electronics standards arena, notably the CAD Framework Initiative, which has based itself in part on the modelling work in EDIF.

The influence of STEP, and particularly EXPRESS, is clear. The idea of views, hierarchies and a formal language predates the idea of application protocols and the STEP work described in Part IV, but has many conceptual similarities. The use of EXPRESS, resolutions to co-operate with the STEP community, and wider common understanding are encouraging developments.

```
edif =
  '(' EDIF ediffilenamedef edifversion ediflevel keywordmap
         { <status> | external | library | design | comment | userdata } ')' .

ediffilenamedef = namedef .

edifversion =
  '(' EDIFVERSION integertoken integertoken integertoken ')' .

ediflevel =
  '(' EDIFLEVEL integertoken ')' .

status =
  '(' STATUS { written | comment | userdata } ')' .

library =
  '(' LIBRARY librarynamedef ediflevel technology
      { <status> | cell | comment | userdata }
  ')' .
```

Figure 6.2 Extract from EDIF syntax.

A technical comparison of IGES and EDIF was performed under the auspices of SERC EASE and an EMR contract in 1988 [38].

Figure 6.2 presents some of the syntax of EDIF, written in an extension of Wirth syntax notation [174]. The EDIF specifications present the syntax in both this form and in LISP. Keywords are shown in upper case and references to other productions in lower case. Any token in angled brackets (< token >) must occur at most once.

ECSI – the European CAD Standardization Initiative – is a not-for-profit organization which aims to promote standards such as EDIF and VHDL throughout Europe. It was incorporated in May 1993; the membership is as follows:

Anacad	Bull	Ericsson	France-Telecom	ICL
Philips	Racal-Redac	SGS	TGI	Thomson-CSF

ECSI receives direct financial support from Europe through the ESPRIT project ESIP. The EDIF Technical Centre at the University of Manchester generously provided the material for this section.

6.2.3 VHDL

In the early 1980s, the US Department of Defense initiated a project to develop a hardware description language. After public review, VHDL became an industry standard [158]. It is used to describe the structural and behavioural characteristics of electronic hardware, ranging from simple logic gates to complex digital systems.

A schematic and high-level behavioural model is required for every level, from the entire system down to individual chips. Each entity must have an associated test bench, so that the design can be simulated. The three main elements of the language are:

(a) design entities, which model hardware of any complexity, and comprise:
(b) interface descriptions, which describe how the entity relates to other entities and the outside world;
(c) body descriptions, which describe how the entity is structured and behaves.

6.3 Open systems interconnection

Product data exchange also has a relation with the open systems interconnection (OSI) model. This addresses the "how" of the communication problem (i.e. it defines the medium). The confusion concerning the distinct roles of product data exchange and OSI is exacerbated by the fact that the top layer in OSI's model is called the application layer.

OSI defines a standard seven-layer reference model (ISO 7498) [105] which specifies the purpose of each layer within the model. The OSI architecture is layered and organized as a hierarchy of independent, but supportive, functional layers. This isolates the complex functions in each layer and also allows each function to be transparent. Each layer encompasses a set of standards which define the services offered as well as the actual protocols to be used.

The layers are as follows:

7 (application): provides all services directly comprehensible to application programs;
6 (presentation): restructures data to and from standardized formats used within the network;
5 (session): synchronizes and manages data;
4 (transport): provides transparent reliable data transfer from end-node to end-node;
3 (network): performs message routing for data transfer between nodes;
2 (data link): detects errors for data moved between nodes;
1 (physical): encodes and physically transfers data between adjacent nodes.

Clearly, the lower the level number, the closer it is to the physical hardware.

A given subset of these standards for the reference model is called a functional profile, which defines a specific OSI implementation. Examples of such profiles are:

(a) Manufacturing Automation Protocol (MAP);
(b) Technical and Office Protocol (TOP);
(c) Government Open System Interconnection Profile (GOSIP).

The user uses the OSI communication system via the application layer which provides a set of accessible services. These are provided by an application service element, such as the manufacturing message specification (MMS). These services are then used without regard for the underlying protocol stack, which performs functions such as session control, error recovery, link control and, of course, data transmission.

6.4 Documents

An important requirement in engineering is the transfer of documents containing both text and drawings. Such documents include product specifications, tenders and contracts. These documents are increasingly produced on word processors or even desktop publishing systems. SGML and ODA are both relevant to how pictures in these documents are created and transferred. Since many of these will be produced by CAD systems, either because they are engineering drawings or because computer-aided draughting systems provide a convenient means of sketching, it is not unreasonable to suggest that IGES would provide a suitable mechanism for transfer. Equally, product data stored electronically may need to reference documents, such as contracts, requirements, national or international laws (such as for safety) with which the product has to comply, and user guides.

SGML (ISO 8879) [139] enables the logical structure and output format of a source document to be transferred between systems. Thus, rather than being a mark-up language, it allows the user to specify the mark-up language to be used. It has two types of declaration. The Document Type Definition defines the organization and logical structure: for example, it indicates how text which is a section heading is identified as such. The Formatting Output Specification Instance defines the format and style: how text marked in a particular way will actually be printed. Consequently, the appearance of a document can be adapted to suit particular local hardware and software requirements without its logical structure and contents being changed. The appearance and content of the document are separated. This is similar in GKS to the definition of output primitives and their attributes, and the separation of logical and physical input devices.

ODA [102] provides an architecture and processing model, rather than being a language. It allows documents to be presented (as a whole) as the author originally intended them to be, irrespective of the facilities of the receiving system. Documents themselves are exchanged using ODIF (Office Document Interchange Format) [103].

SPDL (Standard Page Description Language) (ISO 10180) [144] enables page layouts to be described in a computer-readable form. It accepts output from SGML.

6.5 Business

It was mentioned in Chapter 1 that the commercial ends of manufacturing companies were exchanging information between business partners. There is an international standard for the electronic data interchange for administration, commerce and transport called EDIFACT [14]. ODETTE has defined a format for invoices and orders, and the supporting structure. It is now established and such data is regularly transferred over telecommunications networks. The intention was to do the same with CAD data, a much more difficult operation because of the size and complex data structures required to represent the information adequately.

6.6 Conclusion

The standards and specifications addressed in this chapter are all pertinent to the field of computer-aided engineering in some way. Although none has been described fully, more detailed descriptions of them can be found as follows:

(a) graphics standards [8];
(b) Manufacturing Message Specification [43];
(c) Computer Graphics Metafile [67, 99];
(d) OSI Reference Model, MAP and TOP [41];
(e) structured documents [21]

Lessons were learned from the methods used in the development of these associated standards, and applied in the field of data-exchange standards. A study of the integration of graphics and communications has been produced by Sutcliffe at al. [147].

Part III

Complementary initiatives and infrastructure

As noted in Chapter 3 there were problems with the early second-generation specifications. Chapter 7 classifies these into two types: those which occur generally with product data exchange, and those which are specific to the use of a particular neutral format. Methods for overcoming each type are presented, including mechanisms for improving the specifications themselves.

Various initiatives made in testing vendors' processing software are presented in Chapter 8; and Chapter 9 describes the increasingly sophisticated attempts to produce subsets of a standard, which can be used to support a particular application. Chapter 10 is devoted to the topic of conformance testing, understanding of which grew out of the earlier informal testing and of conformance testing for GKS and OSI.

As with later versions of second-generation specifications, methods for testing and subsetting were defined in parallel with those for STEP. Again, STEP benefited from the practical experience gained with the second-generation specifications which, in turn, had concepts introduced which would be compatible with STEP.

Chapter 7

Problems with product data exchange

In this chapter, the problems with data exchange are addressed by considering first those which occur with all forms of data exchange. Problems which are specific to the use of a neutral format (even one that has been through the standardization process) are then presented, with advice and solutions where possible. Lessons learned by those developing specifications and from the use of the processors in industry are included. As a particular example, the procedures that the IGES committees have initiated for repairing IGES and the concept of recommended practices to help implementors are introduced. The ways in which the use of SET and VDA-FS were promoted provide contrasts from which the IGES committees have learned.

7.1 General data exchange problems

The problems listed under this heading apply to all attempts to exchange information content between dissimilar computer-aided systems, not just those used in engineering. The transfer of documents produced by "intelligent" documentation systems, the transfer of graphical information and the transfer of maps are relatively close examples; the last two are conceptually easier than even computer-aided draughting data; product data exchange is far more complex. The problems are described as follows.

Drawing office conventions
Drawing offices use different conventions to organize the drawings in a single project: for example, how the use of layers is organized or what co-ordinate system is used.

Medium incompatibility

The use of conflicting types of medium such as tape, floppy disk and networks can cause incompatibilities.

Functionally different CAD systems

This covers any mismatch of functionality for which there are not, yet, representations which retain adequate information. The most commonly quoted example is the incompatibility between a two-dimensional draughting system and any system that has as its basic model a three-dimensional representation. Information is irretrievably lost if one translates the three-dimensional model into orthogonal projections in a system where this is the base form. Research is still proceeding on how to interpret three orthogonal views as a three-dimensional model automatically (i.e. by computer algorithms), and this presupposes that the views are consistent [159].

The mismatch may also occur because of the different domains for which each CAD system was designed. Draughting systems for schematics for electrical or hydraulic flow representation have different emphases from those for mechanical CAD. We have already indicated that geometric data for finite-element systems does not represent the complete geometry of a product definition: it has been selected as required for the analysis.

Numerical approximation

Conversions between different surface representations are not always defined and may result in approximation or numerical problems. The same problems can manifest themselves with conversion between different solid representations.

Vendor co-operation

This covers the attitudes of vendors and particularly the availability and quality of the processors they write to support transfer using a particular method. There are further implications in that a CAD system is usually developed in only one country. A slow response to complaints, whether regarding bugs or lack of functionality, can lead to users' frustration.

7.2 Problems with neutral-format transfer

All neutral-format specifications have problems inherent in their design.

Explosion of data

The CAD system implementor can design his data storage according to his priorities and particularly to support the functionality of his CAD system. A format that is going to be general enough to be independent of CAD systems and thus able to represent accurately all the possibilities will naturally be larger. In addition, extra data to allow checks to be made on correct transmission between the systems will lead to an increase in size. Quotations in using IGES range from twice to ten times the original CAD system file size.

Informal specification

Current second-generation formats have been written in English, German or French; that is, in a natural language. Contradictions, omissions and ambiguities have occurred which lead implementors to contradictory interpretations. Such problems also make testing that an implementation fits the specification subject to interpretation; this time the interpretation is made by the tester. A third disadvantage is the difficulty in processing such a specification by a computer to extract the information content, rather than the text that represents it.

An example from IGES occurs in the case where an entity may "use" other entities as subordinates. An offset curve (entity 130) requires another curve from which to construct the offset: ideally an enumerated list of possibilities should be given but in practice a phrase such as "any parametric curve" is used. As one example of a parametric curve is an offset curve, an apparently reasonable option is for the base curve to be another offset curve: doubly offset curves arise as a result. This may indeed have been the intention of the specification's writers, but not all readers may realize such possibilities. Consequently, although some vendors may allow doubly offset curves, others will not. The result is a loss of information when such an exchange is attempted.

Complexity of the format

Engineering information, particularly the inherent relationships between parts, can lead to complex file formats to reflect those relationships that are difficult to understand and implement. Experience with IGES is that this leads to errors and misunderstandings by the implementors. As always, simplicity should be a priority.

Limitation in coverage

It is not realistic to expect a neutral format to cover every aspect supported by all CAD systems. Vendors attempt to include features which will improve the system and make their product saleable by being a step ahead of the competition. This difficulty is associated with "timeliness", as below.

Mismatched subsets

The major problem has been found to be that vendors do not implement all the entities in the specification. Even something as basic as a message from the postprocessor to say that an entity cannot be interpreted in the receiving CAD system is often omitted. No implementation supports every entity defined in a neutral specification, unless (like VDA-FS) its coverage is specific and contains few entities. The postprocessor should inform the receiver about what has not been interpreted at all and about any conversions and/or loss of information that has occurred (e.g. conversion of a spline into line segments). With a specification of the extent of IGES it is unrealistic that a vendor's processor should generate entities that do not occur in the CAD system. Equally, this mismatched subset problem means vendors support different sets of entities and is a frequent cause of exchange difficulties.

Mappings

Implementors must choose the neutral-format entity which most closely matches each of the features in the CAD system. The problem of terminology and the different uses to which the same terms have been put are explained in Chapter 3 where the problems incurred in exchanging between two-dimensional draughting systems were examined. This lack of consistent terminology has led to problems of interpreting the product data exchange specifications. For example, in IGES, should a symbol be represented using the subfigure entity (308), one of the group (402) entities, or as a general symbol (228; available in IGES 3.0)? It has also been found that a straightforward CAD system circle has been represented as a composite curve (102), a conic (104), and a rational B-spline curve (126), as well as the natural choice of a circular arc (100). In some systems, circular fillets are written either as circular arcs or as conics, depending on whether they were defined initially by radius or by centre. The choice of which mapping to use for each CAD system feature in the preprocessor but also in the postprocessor has a tremendous effect on the quality of data exchange that is possible.

Timeliness

Neutral formats attempt to support existing CAD systems; in the case of IGES 1.0 in 1980, this was the most sophisticated of the computer-aided draughting systems then in use. Neutral formats take time to go through the standardization process which is why people start experimenting with them before standardization. Though the experimenters might spend time finding errors in the specification, or in implementing constructs which will change, they can also provide valuable input to the standard makers. Also, time is a reason why some people ignore standardization: this time factor means that the standard, when it appears, is often behind current computer usage, a common problem with information technology standards in general.

7.3 Addressing the general problems

Much effort has been contributed to addressing the problems listed above in an attempt to make data exchange reliable and robust. This has resulted in guidelines and recommended procedures for the users, more help for the implementors, and improvements to specifications and the support of those specifications. The problems are now addressed in order.

Drawing-office conventions
This is a problem which can only be solved by the co-operating parties. Their attention is drawn to potential problems by guidelines such as those produced by NEDO [16] in which there is a section devoted to the drawing-office incompatibilities found in the study.

Medium incompatibility
The major problem reported at the first public demonstration of IGES at Autofact 1983 concerned the reading of magnetic tapes. The testing done by the NEDO group at about the same time highlighted similar problems to such an extent that their guidelines included recommendations for magnetic tape usage. This included the type of tape, the formatting of the tape and the handling thereof. It is salutary to think this occurred some ten years after the standard for industry-compatible magnetic tape had been ratified [6].

In the early 1980s, magnetic tape was almost exclusively the medium used for transfer. It is still common. With the increasing use of CAD systems based on personal computers, particularly in the construction industry, floppy disks are being used for exchange. The CADCAM Data Exchange Technical Centre has updated the NEDO guidelines to include the IGES recommended practice (RP6) and suggestions for the use of floppy disks.

This latter also looks ahead to the transmission of neutral-format files over networks and the CADDETC guidelines contain some preliminary suggestions. TOP (see Ch. 6), a suggested implementation of OSI for technical-office transfer, includes a subset of IGES entities in its application layer (layer 7 or 8) as an example of the type of information a technical office would transfer over a network. Except in the most advanced companies it is felt that it will be several years before transmission over networks is commonplace and one has to be optimistic that the difficulties we currently face will have yielded to the onslaught of open systems (i.e.

widely used standards supported by formal testing to engender confidence and thus use).

Functionally different CAD systems

It is always going to be true that CAD systems and their more intelligent successors will exhibit different functionalities. One can only understand the differences and then consider whether it is worthwhile to undertake a transfer that causes loss of information or involves interpolation to create the data that the receiving system demands. The future envisaged by STEP will have a product model which can hold all the life-cycle data and therefore in theory support all functionality. The original aim was to cater for all industrial sectors in the same framework. The understanding required to achieve these aims is not yet available.

Numerical approximation

In the early days of IGES, the National Bureau of Standards provided algorithms coded in Pascal to transfer between curve and surface representations: specifically, between the rational B-spline (126 and 128) and parametric (112 and 114) curve and surface forms [54]. Research into accurate transfer between different polynomial forms and polynomials of different degree is continuing [37], along with research into methods of checking that two representations represent the same surface [22].

Newer CAD systems are using the non-uniform rational B-spline (NURB) representation because a wider variety of curves and surfaces including conics can be represented exactly. This may alleviate conversion problems in the future. Indeed, some companies have suggested that the NURB should be the curve and surface form into which all others are converted (i.e. it should be the only form in the neutral format specification or be used as the database archiving form inside a company).

Vendor co-operation

In the early 1980s, CAD vendors and their representatives were the source of many of the problems because processors were inadequate. As in all software there were bugs, they were not robust, and they failed without explanation when they met entities defined in the specification but for which provision had not been made in the software.

In addition, the vendors often did not give the impression that they wished to sell the idea of neutral format or indeed any other type of data exchange, except to receive data into their own system. Intergraph and IBM produced their own formats to encourage this (see Ch. 2). One can sympathize. If a vendor is able to sell his system and "hook" the user so he finds the cost of changing to another CAD system prohibitive it is in the vendor's interests to do so. In computer procurement this is the norm. Any mechanism which allows data exchange (and particularly the archiving of data that will be accessed by some as yet unknown future system) is undesirable. Indeed, as with the contractor/supplier relationship, it is the smaller vendors who have supported exchange mechanisms in order to use them as a selling point.

The education of users and the support given by the IGES committees have improved processors and the increasingly competitive CAD market has perhaps changed vendor attitudes. The publicity given to open systems and the frustration caused by communications problems at all levels of computer usage impacts on an increasing proportion of the population and should lead to change: user demand for reliable, easy communication.

The incorporation of a standard product model as the basis for all CAx systems is a long way off and sales are made by having unique required features. However, there may come a time when a selling feature is that one's system is based on a standard product data model.

Software support can only be solved by user pressure which leads the vendor to improved software development methods and a willingness to submit his processors to testing. In general, vendors have been supportive of the testing that has been pursued extensively although informally during the 1980s. The need for more formal test procedures that can rely on well specified standards is shown in the next chapter which identifies the lessons learned from all the testing activities. The problem with neutral-format processors has been differentiating between poor software, inadequate interpretation of the requirements, an ambiguous specification of the standard, and lack of coverage. Education in all its forms is the only way to enable users to make the correct interpretation of the cause of their problem and thus exert pressure in the most appropriate form.

7.4 Addressing the problems inherent in a neutral format

We now concentrate on problems inherent in the use of neutral formats. These are largely the concern of the standards makers: not just in developing the appropriate functionality (i.e. support of adequate information content), but also in ensuring that there are mechanisms and methods to make it easy to write good processors. End-users need to be convinced that the process is reliable and easy.

Explosion of data

The solution to this problem lies with the specification makers and although the authors do not feel this is the most serious problem, much was made of the size of IGES files in its early days. The SET development was initiated to specify a more economical format and files of a tenth the size of those of IGES have been quoted. The IGES committees responded to the criticism, first by defining a binary form in IGES 2.0 and then, in IGES 3.0, introducing a compressed ASCII format which eliminated insignificant spaces and reduced the size of the directory entry and parameter data sections. The binary format was deprecated in IGES 5.0. (The reduction in hardware costs and the provision of operating system "compress" utilities have alleviated this problem.)

The lessons have been learned and are incorporated in the STEP physical file format which was initially proposed by a small team which included the SET and CAD*I developers and the PDDI participants.

There are places in the IGES syntax where a particular number in the file dictates what follows. The development of a context-free syntax – that is, one which can always be parsed with no knowledge of the particular instance – would also facilitate the writing of processors.

Informal specification

Again, the form of a specification is the responsibility of the standards makers and the need for a formal specification language suitable for reading by a computer and a human was realized by the developers of EDIF and CAD*I. A formal language has a well defined syntax, which is itself defined using a notation such as Backus–Naur Form [101] or WSN [174]. This data definition language must also be capable of representing in as readable a way as possible the information structure including relationships between entities (using that word loosely at the moment) and constraints on their use. An example of a representation of syntax with explanation is shown in Appendix Figure C.3. Clearly, there must be software to check that code in the language conforms to the syntax and also to compile it into some form that can be used for further code generation.

To overcome the deficiencies in second-generation neutral formats which affect the quality of processors, all that can be done is to provide guidelines and ensure that revisions of the specification strengthen definitions where ambiguities have been identified. The mechanisms set up by the IGES committees are discussed later in this chapter. The French have an advantage in that an influential user promoted SET and helped vendors such as Computervision and Dassault (the French aerospace manufacturer that was the originator of IBM's Catia) to implement SET processors. This almost paternal interest and the fact that SET has a simpler file structure have reduced problems in this area. However, it should not be imagined that all is rosy in France: visitors to French CAD conferences hear the same complaints about neutral formats not working as are heard elsewhere. There is now GOSET, which is still largely made up of SET's original developers but providing a service to all sectors.

To address the problem of options available within a particular entity, later versions of IGES are introducing conformance requirements, which enumerate the possibilities that may be used in a given context.

Complexity of the format

This again is a problem that has to be solved by the specifiers. The information to be transferred is complex. Current database technology does not adequately represent all the relationships and constraints we would wish for product definition data, let alone product data in its entirety. Clarke et al. [29] discuss the requirements for representation and their report was commissioned to investigate whether object-oriented paradigms were more appropriate and available. At that time the conclusion was that object-oriented databases were not ready, although they probably were better than older paradigms.

Thus, how to represent the data is not yet agreed, partially because of its complexity. There are two issues: the complexity of the information and the size and complexity of the physical file format that is used to transfer the data. IGES has a complex file structure, which is also large. SET shows that the same information can be represented more simply. The STEP developers are attacking this issue by considering the different levels of representation and have proposed a three-tier architecture based on ANSI/SPARC database proposals [150].

The effect of complexity on the use of current-generation formats can only be ameliorated by providing help for implementors as described above. The establishment of advice centres such as CADDETC and GOSET has proliferated throughout the industrial world and is the common response to all the neutral-format problems that cannot fully be rectified without an entirely new specification.

Limitation in coverage

Vendors can encounter the problem that some of the functionality in their CAD system is not covered by a neutral format. One example is the inability to transfer parametric representations completely. What should the vendor and user of such a system do? The user needs guidelines from the vendor about exactly what can be exported and received; if the coverage is adequate for the receiving system, the sender can impose conventions on the use of the CAD system; however, these may be unduly restrictive. Part of the NEDO guidelines concerned drawing-office practice, which included understanding the capabilities of both sending and receiving systems to identify mismatches. Some of this is caused because the format has limited coverage. It is reported that CADCentre, the Cambridge originators of PDMS (Plant Design and Modelling System) used in the process industry, did not implement IGES 1.0 and 2.0 processors because their system in its coverage of three dimensions, and particularly the relationships in the bill of materials, was in advance of the specification. Clearly, IGES would have been better sooner in the AEC area had such understanding been contributed at the beginning, rather than later, but this would have resulted in commercial competitors also gaining the benefits.

This limitation in coverage is the converse of the next problem, mismatched subsets, but will be more difficult to solve as it would require a perfectly comprehensive specification. This is STEP's intention but is not currently achievable. There may come a time when vendors limit their capability and expression of it to that in the standard, but it is too early to expect or require that yet.

Mismatched subsets

This is the most widespread source of exchange difficulty but one that does appear amenable to solution. That it was not foreseen is a lack in the original work on IGES, although it is easy to be wise after the event. GKS (see Ch. 6) [56] which came somewhat later, introduced the idea of subsets and these have been comprehensively adopted by IGES supporters, including the IGES committees and influential bodies like the US military and the association of German car manufacturers (VDA). The influence of this on the content of STEP is so important that the whole of Chapter 9 is devoted to the development from subsets to application protocols in STEP and the help being given to current users.

Advice and guidelines on use of specifications and information on the coverage of individual CAD systems will enable users to overcome this problem. However, although informal subsetting guidelines may be sufficient to make data exchange more reliable in an interim period, the error in the standard which caused such guidelines to be introduced still needs to be rectified.

Mappings

Without guidance, implementors made many apparently inexplicable decisions on how to map a CAD system entity onto a neutral-format entity. Some of this was due to inadequate explanation of intention in IGES. An example is lack of explanation of the purpose of the IGES group entity with its many forms compared with that of the subfigure. The problem is such that implementors and those who test processors have to make assumptions because different CAD experience can lead one to misunderstand the purpose of IGES. The real solution is to develop better formulated neutral formats which incorporate in their formality prescriptions for use. As described in Chapter 9, such prescriptions are an important part of the German response to subsetting: they detail how the entities should be used and prescribe what conversions should be made where these are necessary.

Timeliness

The problem of the time it takes to make a standard is not simply soluble. There are moves to speed up procedures in light of the difficulty users of information technology are finding. There are always conflicts between those who want a working solution immediately and those with no immediate financial pressures who are more interested in a longer-term solution. It is not surprising to those involved that STEP was expected in 1987 and is only now beginning to be used, in the mid-1990s. This does leave the user community with problems. The authors do not feel that large reductions in time are possible if the result is to be useful. However, understanding of the process and the reasons for the formality are important.

7.5 IGES studies

Interest in data exchange and IGES in particular started to spread in the early 1980s. Two particular activities were undertaken which were relatively independent of the NBS-led IGES work in the United States. They are notable largely because they demonstrate the impact of IGES at that time and the general lack of understanding of the scale of the issues that needed addressing to enable the complete information exchange which was required.

The NEDO report [16] is entitled *Computer interchange of data: guidelines for the use of IGES. Phase 1: two-dimensional scale and non-scale drawings with text.* The work was instigated by the CAD group of the Engineering Construction and Process Plant Economic Development Committees (EDC) in the United Kingdom. EDCs ran under and were serviced by the National Economic Development Office which promoted contact between British government, industry and trade unions. This CAD group under the chairmanship of Roger Dobson of Bechtel (GB) Ltd, a multinational construction company, was established in 1983 with membership from clients, contractors, manufacturers and vendors. It was instructed to examine the problems of exchanging data in a machine-readable form between CADCAM systems from different vendors and to recommend methods to achieve this objective.

The group examined the various options and decided that two-dimensional drawings, which formed the major portion of communications in their industries at that time, would benefit from the use of IGES. It then set out to test transfers with the willing co-operation of companies in their sector, involving CAD vendors to provide information about their IGES support.

The work was executed with the help of the Department of Mechanical Engineering at the University of Leeds, which had funding from the Department of Trade and Industry for six months. The results of testing were so dire that there was some discussion about the form of the report. It was published in May 1984 and indicated the potential problems in CAD data exchange such as system mismatch and progressive loss of information. It attempted to describe accurately but impartially the current capability of IGES, the availability of processors and their (lack of) quality, and also collaborative initiatives and interest groups involved with improving data exchange. Many of the initiatives described earlier in this book were mentioned in the NEDO report.

The major portions were guidelines for managing exchange and also procedures to be adopted by those using CAD systems in the drawing office. Suggestions were made on the use of magnetic tape and how to go about testing an exchange before committing valuable data to the process. The preface to the report says it all:

> . . . although the results of testing have not been satisfactory yet. This failure is not due to the failure of IGES or the guidelines, but to the fact that the majority of the vendors have not fully implemented IGES, nor have given this implementation their wholehearted support. This will only be changed by pressure from the users. In order to generate this pressure, users must know what is possible and its value to their operations. This document is published in the belief that it can provide the necessary education and thus stimulate user action.
>
> At the start of the work, the NEDO CAD group advised the authors on a pragmatic approach to finding the best available mechanism for interchange of drawings in computer-readable form. The view then, which has been confirmed by the study, was that IGES . . . was the data exchange format on which efforts should be concentrated. Those who have studied IGES appreciate the technical expertise which has defined the format, given the aim of CADCAM system independence. There are well advanced programmes to address the omissions and limitations. It is hoped that this report will encourage British contributions to the IGES Revision Committees and other forums.

Although today that may appear somewhat naïve, the overwhelming badness of the processors of even the largest companies who were working in IGES committees and the common comment we heard, "IGES is useless", will have influenced us.

The report was widely read and did provoke activity and comment. A second version, which extended the domain of interest to the additional data associated with drawings and often presented in tabular form and which attempted to improve the presentation and update the report, was published in July 1986. The work led to the establishment of CADDETC (the CADCAM Data Exchange Technical Centre) at Leeds to help UK industry with its exchange problems. Work has thus been able to continue on improving and promoting data exchange methods.

The Booz, Allen and Hamilton IGES [19] report was produced as the first task of the PDDI project, and compared IGES with PDDI, as well as enumerating problems that could occur with the use of IGES.

AECMA is the European aircraft manufacturers association. It produced a report [1] which compared IGES and SET for a typical drawing. One of the most memorable results is a before-and-after picture of a drawing: the latter was very tiny with apparently very little information on it. The scaling technique of the postprocessor was a little eccentric and size had not been properly accommodated, a common fault in early transfers but one which was easily rectifiable.

7.6 Lessons learned

Second-generation transfer methods have been used to refine the requirements for STEP. The important factors are: better methods for defining the specification; fuller, more complete, coverage of product definition data; and formal testing to encourage better processors and thus user confidence.

7.6.1 Definition methods

Rather than use a natural language exclusively, more formal definition methods are available. These include:

(a) the language used for definition of the standard (a data definition language);
(b) the form of the standard so that it can handle the complexity and amount of product data, both for transfer by computer and for human interpretation;
(c) an information modelling language to enable the information that practising engineers know about to be communicated as requirements to the standards makers for formal definition;
(d) recommendations for how the parts of the standard should be used, including constraints and mappings.

7.6.2 Coverage

The extent of the specification is a largely unsolved problem. Second-generation neutral formats covered product definition data, as manifested on engineering drawings. There were some extensions which reflected additional facilities vendors had added to their computer-aided draughting systems, such as three-dimensional wireframe modelling, which allowed computer generation of perspective views. The need to represent surfaces and to support finite elements reflected the islands of automation that had grown up. For mechanical CAD there is some use of solid modelling but our understanding is not yet adequate to define a system-independent neutral format that supports all forms of computer representation of solids. For electrical and process industries, the drawing of schematics and the connectivity they embody can be supported. Even EDIF, for all its advanced implementation technology and intent to be a design format, is widely used for printed-circuit-board schematics.

STEP was intended to cover all domains in a single methodology and framework. The initial release of STEP [172] was intended to cover only the same functionality and domain as IGES did at that time, although IGES had, built-in from its inception, associativity mechanisms which allowed projections to be linked to a model, or text to geometry. It is the methodology and the framework, which it is believed are extendable, that will be new. The understanding of the form and structure of product data even in mechanical engineering is an active research field, not one that is ready for standardization.

7.6.3 Formal testing

The more formal definition methods facilitate the documentation of the requirements of a processor. In addition, the availability of machine-readable specifications allows some automatic code generation: software can parse parts of the specification which is to be supported. Even so, it is still necessary to test the software.

Testing that software conforms to a standard is not new. There are testing services for language compilers (e.g. Fortran, Cobol and Ada). The amount of testing that IGES and other second-generation formats encouraged and the parallel requirements for OSI networking components has led to investment in the establishment of testing methodologies and services using them. Early learning from these activities emphasized the need for formally defined specifications and completeness criteria that application protocols in STEP provide. As described in the next chapter, conformance testing is only a start to ensuring that processors are reliable enough for user confidence.

7.7 IGES-centred improvements

7.7.1 Recommended Practices

After the appearance of IGES Version 3.0, the IGES committees instituted a series of documents called Recommended Practices [10]. These clarify some of the ambiguities and misunderstandings in IGES, without having to wait for the next version of the specification to be published; indeed, some of the Recommended Practices have been superseded in a later

Table 7.1 IGES Recommended Practices.

	Recommended Practice	IGES version	Processor pre-	Processor post-	Applicability to IGES construct
1	delimeter †	3.0 (4.0)	✓	✓	GS1 GS2
2	witness line suppression	3.0	✓		106/40 DE9
3	transformation of nested subfigures †	3.0 (4.0)	✓	✓	308 408 DE7
4	transformation matrix processing	3.0	✓		124 DE7
5	system id parameter †	3.0 (4.0)	✓		GS5
6	magnetic tape format	3.0	✓	✓	none
7	maximum co-ordinate value	3.0	✓		GS20
8	independent witness lines	3.0	✓	✓	106/40 DE5 DE9
9	postprocessor diagnostics	3.0		✓	none
10	text font 0	3.0	✓	✓	212
11	user interface	3.0	✓	✓	none
12	flag note restrictions †	3.0 (4.0)	✓		208
13	views for transformations	3.0		✓	124 (410)
14	translator documentation	3.0	✓	✓	none
15	zero radius arcs	3.0	✓	✓	100
16	translation vector	3.0		✓	124
17	model space scale	3.0	✓	✓	GS13
18	processing macro entities †	3.0 (4.0)	✓	✓	306 600+
19	independent and dependent processing	3.0	✓	✓	DE9
20	back pointers in view associativity	3.0 (4.0) (5.0)	✓	✓	402/3 402/4
21	comments in parameter data records	3.0	✓		all DE14
22	bounded planes †	3.0 (4.0)		✓	108
23	ruled surface discontinuity	3.0	✓	✓	102 118
24	representation of linear strings	3.0	✓	✓	100 102 106 110 112 114 126 128
25	minimum user-intended resolution	3.0	✓	✓	GS19
26	arrowhead and leader line data	3.0	✓	✓	214 106/40
27	limitations on level identifiers	3.0	✓	✓	all DE5
28	single/double pass processing	3.0		✓	all
29	spline curves and surfaces	3.0	✓		112 114 126 128
30	global parameter defaults	3.0	✓	✓	GS (all)
31	hierarchy flag †	3.0 (5.0)	✓		all DE9
32	arrow and witness pointers	4.0	✓		202 216
33	entity origins	4.0	✓		124 212 312 404 408 410
34	text rotation angle	4.0	✓		212 312 (124)
35	duplicate transformation matrices	4.0	✓		124 (104)
36	excess characters in real numbers	4.0	✓		all
37	reduce i/o overhead	4.0		✓	all
38	unique subfigure names	5.0	✓	✓	308 320
39	empty subfigures used as external references	5.0	✓	✓	308 320
40	scale in matrix	5.0	✓		124 (408) (410)
41	entity identifiers	5.0	✓	✓	308 320; all DE18 DE19; 406/15
42	unique names	5.0		✓	308 320 406/15
43	closed areas	5.0	✓		106/11 106/63 230
44	text box processing	5.0		✓	212
45	null text string	5.0	✓	✓	212
46	drafting symbols	5.0	✓		212 228
47	zero area surfaces	5.1	✓	✓	114 118 120 122 128 140 143 144 108
48	generic data property	5.1	✓	✓	406/27 (322) (422)
49	multiple direction entities	5.1	✓		123 190 192 194 196 198
50	multiple occurrences of properties	5.1	✓	✓	all 406
51	loops and faces	5.1	✓	✓	508 510 514 186
52	drawing entities that contain no views	5.1	✓	✓	404 410

IGES versions in parentheses indicate the version at which a Recommended Practice became superseded (made obsolete or deleted); the RP is also marked with a †.

Problem statement:

There is no standard for the exchange of IGES data on physical media such as magnetic tape.

Alternatives examined:

Media for data transfer can be agreed upon by the sender and receiver of the data. The media consists of magnetic tape, floppy diskettes, or telecommunications.

Selected methodology:

Use magnetic tape with the following characteristics based on ANSI X3.39-1973 and ANSI X3.22-1973:

- one-half inch tape
- unlabelled
- 1600 BPI density
- 9 tracks
- 7-bit ASCII code (8th bit zero)
- 80-character fixed length records
- 10 records per tape block (this implies a blocksize of 800; 80 character length * 10 records)
- a single end-of-file mark separates IGES data sets on tape
- multiple end-of-file marks signal the end of tape data

Figure 7.1 IGES RP6.

RP 4 problem statement:

The identity matrix in field 7 of a DE record can be identified in several ways.

Alternatives examined:

1. Use a pointer to an identity matrix defined as an Entity type 124.
2. Use the "default" value of 0.

Selected methodology:

For processing efficiency, it is preferable to use a zero in field 7 of the DE rather than to use a pointer to an identity matrix.

RP 35 problem statement:

Some preprocessors associate a Transformation Matrix (124) with every Circular Arc (100) even when they contain identical data. For example, holes on a flat surface have a common normal vector, but the preprocessor gives each circle its own (identical) 124 instead of pointing at a common one. Another problem is that the preprocessor may set the origin in the parameter data section to (0, 0) and generate 124s with identical rotation matrices but unique translation vectors.

In a slightly different vein, some preprocessors create Circular Arc entities such that the start angle is always zero (Y2=Y1), and create a Transformation Matrix which rotates it into position. In the case of filleted corners of a cut-out on a planar surface, they could all use the same Transformation Matrix if their start and end points are adjusted.

Either of these implementations leads to needlessly large IGES files, and places a processing burden on the postprocessors.

Although the circular arc is used to illustrate the problem it also occurs in matrices associated with Subfigure Instances (408), dimension entities (200 series), and other geometric entities.

Selected methodology:

The preprocessor should keep track of the Transformation Matrices that it has generated and point to pre-existing ones instead of creating new ones.

However, there are legitimate reasons for creating matrices with a common normal vector; Conic Arc (104) entities on the same plane but with different rotation angles are one example.

Figure 7.2 IGES RPs 4 and 35.

version of IGES. A complete list of IGES Recommended Practices is given in Table 7.1. Each consists of a problem statement, the solution, the rationale, the alternatives considered and the selective methodology. RP 6 (magnetic-tape format) and RPs 4 and 35 (duplicate transformation matrices) constitute Figures 7.1 and 7.2.

7.7.2 Changing IGES

There is a well defined procedure for changing IGES. Anyone who can identify a problem or suggest an addition completes a Request for Change (RFC) form, shown in Figure 7.3. The suggestion will be reviewed by the IGES RFC Committee which will evaluate the impact of the change and, if it is judged worthwhile and feasible, pass it forward for voting by the General Assembly. A general ballot follows. If it is accepted, an Edit Change Order is produced, which gives precise instructions to the editor of the IGES specification. This will be effected in the next version of the specification. Proposed Recommended Practices go through a similar procedure. One can understand that people may become frustrated with the whole standards-making bureaucracy, because of the time taken to correct a problem, but it is essential that only considered changes are made to a specification of which implementations exist already.

7.8 Software for testing

Several software tools exist for checking the files produced by neutral formats, in that they should conform to the specification. These tools are primarily for the use of end-users wishing to exchange data. We thus differentiate them from processor testing which is a more complex procedure addressed in the next chapter. That is not to say that some of the tools cannot be used for both kinds of testing.

Such tools include the following.

Syntax checkers for diagnosis

These were the first to appear, both in the United States and in Europe. There are examples that date from 1983 and they test the file format, syntax and semantics. An additional check which has become increasingly important is that of conformance to some given subset (whether it is widely accepted, such as VDA-IS [155] or one of the classes in MIL-D-28000A [97], or defined by the user).

Transfer forecasters

From the early days potential users wished to know the likely success of a planned transfer between two systems or to know the likely

IGES
REQUEST FOR CHANGE

Pg____of____

NAME	DATE	SERIAL NO.

MAILING ADDRESS

PHONE EXT.

CHANGE TYPE

☐ ERROR CORRECTION

☐ CLARIFICATION

☐ EXTENSION

SUBCOMMITTEE APPROVAL

CHAIR SIGNATURE

DATE

REVISIONS			
LTR	DATE	AUTHOR	DISPOSITION

DESCRIPTION OF PROBLEM

SOLUTION(S) PROPOSED

Figure 7.3 IGES Request for Change form.

consequences of a planned change of CAD system. Reports by NEDO [16, 17] and Booz et al. [19] attempted to help by listing CAD system preprocessor and postprocessor coverage. Today, a forecaster will cover a range of systems and neutral formats and will have gathered its information from vendors, from user experience and by both informal and formal testing. Obviously, the value of such a database depends on constant access to up-to-date information from all sources.

Editors for changing files

The experience of data exchange throughout the 1980s has been that neutral-format files have been incorrect on receipt and it has become commonplace to make changes by hand. This practice should not be allowed to continue, especially where contractual responsibility might be impaired, such as in conditions where one would not change a drawing after it had been received. However, software is appearing and has been used to "flavour" IGES files, an increasingly used and accepted activity which involves representing IGES entities in alternative forms or combinations which are acceptable to the receiving system. Some sending systems prompt the user for the details of the receiving system, so that the IGES can be flavoured during its generation.

Some of the earliest flavouring work was undertaken by the DOEDEF project in the USA [5] in order to overcome problems of poor IGES implementation and mapping choices. Other problems which the project identified were system differences and user conventions; this is consistent with the description in Chapter 3.

File generators for testing.

One of the problems with early informal testing was obtaining correct files. Hand-coded IGES files were notoriously poor, because of the built-in redundancy of the specification. Files supplied by particular vendors would invariably contain unwanted entities, the removal of which had to be undertaken by hand. With the pointers between the directory entry and parameter data sections, this is difficult to perform correctly by hand. Consequently, file-generation software which was independent of all CAD systems became available.

Other software has been written that is more appropriate to more formal testing than user planning; this is discussed in later chapters.

7.9 Exchange agreements

Clearly, two co-operating partners wishing to exchange data between their systems must undertake considerable planning. An exchange agreement that might be the culmination of that planning and whose contents act as an *aide-mémoire* for issues to be considered has been drawn up by CADDETC. Drawing-office practice, hardware considerations, transfer forecasts and trial exchanges using small but representative datasets should all be considered as prerequisites for exchanging large volumes of production data.

7.10 Conclusion

Use of the second-generation neutral formats showed that some of the problems encountered were specific to them, whereas others applied to data exchange in general. Many of these could be solved by improving the specifications and by holding discussions between the sender and receiver before exchanging such data. Users began to develop tests to check that the processors they used could handle a particular feature in a specification, and software to check the efficacy of the files they received. Such tests and software were used in testing initiatives, which are described in the next chapter.

Chapter 8

Testing

That any product must be tested before it is used in production is accepted throughout engineering. In general this will be undertaken before the product is put on the market. However, software has been notorious for being released before adequate trials have been undertaken as judged by its performance in use. A problem with all software, but particularly that to support CAD data exchange, is that the requirements have not been sufficiently clear. One reason for this in CAD data-exchange processors is deficiencies in the specification of the formats. As a consequence of all this, widespread testing became the norm.

A testing organization may be one of three types:

(a) that which performs in-house testing for vendors;

(b) that which tests a product after its release either as part of a service to users or as part of the planning and quality checking before and during an exchange between co-operating companies;

(c) an independent laboratory which provides testing services, to ensure that products perform as well as possible and to enable the user to choose products which have a certification mark (in the UK, this is the "kitemark").

The amount of testing in the second category for IGES was (and still is) large. Some of the initiatives and the lessons learned are discussed below.

It should also be pointed out that successful transfer depends on two independent pieces of software written by two different vendors. The user's priority is that a particular pair of preprocessor and postprocessor, each supplied by a different vendor, work together to effect an exchange. A vendor requires that his preprocessor and postprocessor work effectively as two independent pieces of software so the pertinent tests are from CAD system to neutral format for the preprocessor and from neutral format to CAD system for the postprocessor. These two different emphases are reflected in testing procedures.

As shown in Figure 2.3 four processors are needed for data to be exchanged between two systems in both directions. Each vendor is concerned with either the left-hand or the right-hand pair, whereas each user addresses problems with the upper or lower pair.

8.1 Autofact

The early Autofact demonstrations were precisely that: demonstrations that the data exchange using IGES could work. STEP demonstrations, in the same spirit as those IGES demonstrations, have already been given, even though the standard has only just been published as an International Standard. The Autofact work was part of the promotional activities of the IGES committees and is fully described in Chapter 3. Participants were presented in good time with a part to be exchanged and had to demonstrate that they could produce IGES files that could be read by a second vendor, producing a drawing as evidence. These drawings were compared by eye. Software for IGES file comparison (as opposed to file comparison on a character-by-character basis) was not then available.

This sort of demonstration is valuable, particularly as it is now well understood that any program of exchanges requires friendly cooperation between the parties and will not become "black-box" or "appliance" technology in the foreseeable future.

8.2 The NEDO program

An aim of the NEDO study at its start in 1983 was to report on vendor support of IGES with a view to helping users. At the end of Phase 1, which lasted six months, tests had not been completed even though vendors in Britain and users had been helpful. The vendors chosen were those with a presence in Britain and the support was sought from the people on the ground in Britain. One of the complaints is still common: that software support by sales staff in Europe does not compare with that in the USA where development takes place (or, at least, that is our perception from this side of the Atlantic).

The report presented test procedures as guidelines for IGES users. This was the first attempt to formalize such procedures, although the IGES Test, Evaluation and Support Committee had developed, documented and made available several IGES test files. Its original intent was to guide helpers in the envisaged testing for NEDO but it proved useful for end-users. CADDETC based an informal test program on these procedures.

The intent of the testing was to ensure no vendor bias and thus the test data had to be obtained impartially by deciding on the IGES entities to be tested and issuing test drawings containing these entities to all parties. The test comprised:

(a) an unscaled two-dimensional schematic with border and subfigures;

(b) a scaled two-dimensional drawing with annotation (text, dimensions, symbols) obtained from a base three-dimensional model.

The drawings could only contain entities relevant to Phase 1.

Both drawings were circulated to the participants with instructions on how each user should create the drawings in his CAD system. He had to create a plot of his CAD system representation and indicate where adjustments were made to mimic the original drawing more closely – still using the agreed IGES set. IGES files were then to be produced.

A reflection test was suggested: that is, reading the IGES file back into the CAD system which had created it and producing a drawing. The intent was to test if the preprocessors and postprocessors were symmetric.

The third part of the procedure was called the transmission test. A pair of vendor systems was tested by members sitting down together and planning the exchange using the guidelines produced by NEDO. Each tester was to act as both sender and receiver with the requirements of the receiving system taking precedence over those of the sender. It was realized that it may be necessary to modify the source tape in order for the receiver to accept it. The transmission was deemed successful when both IGES tapes were received into the receiving system with a drawing style and intelligence that could be processed by that system.

The recording of alterations and failures was also required. It is not surprising that in the timescale of Phase 1 no testing was achieved and in Phase 2 errors occurred so frequently that it was rarely worthwhile or feasible to pursue testing to its conclusion. Interest in these early tests is more to show how they foreshadowed and how their intention can be mirrored in current testing methods submitted for accreditation.

The outcome of the testing was presented in the Phase 2 report. Applicon Bravo, IBM CADAM, Computervision CADDS4, Pafec DOGS, Ferranti (now Infographics) CAM-X and CIS MEDUSA were variously tested by vendors and users. The results went into the CADDETC forecaster software and the NEDO report [16], thereby providing a general understanding of problems with data exchange. Scale, file sizes larger than expected even for IGES, difficulties with handling symbols and text, and the inability to make good use of the start section were common difficulties.

8.3 NAVFAC tests

In 1986, the United States Naval Facilities Engineering Command (NAVFAC) made public some IGES testing it had instituted. Its aim was to ensure that suppliers (i.e. those companies from whom it was to receive drawings) could produce them in a way such that the NAVFAC system could receive them. They created a test drawing which contained examples of all the IGES entities. Contained in the drawing, shown in Figure 8.1, are all the instructions as IGES text. They instructed that drawings should not be mimicked; that is, they should not be simplified by splitting into IGES entities different from the one suggested in the text. For example, the B-spline was to be input using a B-spline function in the CAD system rather than another curve or curves, where "curve" at its simplest includes straight lines. However, the information provided on the drawing as instructions for input often relied on the creation of a shape that was similar visually, rather than providing actual numerical values.

The resulting IGES files were then postprocessed into the Computervision system at NAVFAC. Many companies attempted to enter the drawing. Their efforts highlighted the lack of control the user has over the eventual IGES form into which the preprocessor will translate his CAD system entity. He may well use a B-spline within the CAD system but the vendors may choose to translate it into IGES in some simpler form. The vendor is not entering into the spirit of IGES if he chooses to lose information in the translation from his CAD system into IGES. This impacts directly on conformance testing methodology and the need to ensure just what a particular system can support.

8.4 Informal testing by CADDETC

One of the reasons for which CADDETC was established in 1985 was to perform validation of IGES processors. Initially at least, the precise meaning of the term "validation" was unclear; it was interpreted by CADDETC as completion of the testing initiated by NEDO, followed by increasing formality supported by software to help in the process.

A programme of testing preprocessors and postprocessors was initiated using the quality-control, file-syntax and semantic-checking software of Lucas. The testing was of processors which supported IGES 3.0 and is documented in a series of reports; the coverage is shown in Table 8.1. The testing was phased, based on the first proposed subsets for IGES, which were also convenient for testing, and took place between 1987 and 1989. Member companies did the testing on their in-house CAD systems. The need for software to generate IGES files became apparent: Luigi was developed at CADDETC to fulfil this requirement. Evolving

Table 8.1 CADDETC informal IGES testing programme.

Test	CAD system									
	Anvil	Bravo	Cadam	DDM	Cadds4x	Dogs	Duct	Graf-tek	Inter-graph	Medusa	Uni-graphics
NAVFAC			✓	✓	✓	✓		✓	✓	✓	✓
g1 pre			✓	✓	✓	✓	✓	✓	✓	✓	✓
g1 post	✓	✓	✓	✓	✓	✓	✓	✓	✓	✓	✓
global pre	✓	✓	✓	✓	✓	✓	✓		✓	✓	✓
global post	✓	✓	✓	✓	✓	✓	✓		✓	✓	✓
a1-3 pre	✓	✓	✓	✓	✓	✓	✓	✓	✓	✓	✓
a1-3 post	✓		✓	✓	✓	✓	✓	✓	✓	✓	✓
g2g3 post	✓	✓	✓	✓	✓	✓	✓	✓	✓	✓	✓
g4 post	✓	✓	✓	✓	✓	✓	✓	✓	✓	✓	✓
s1 pre	✓	✓		✓	✓	✓	✓	✓		✓	
s1 post	✓	✓	✓	✓	✓	✓	✓		✓	✓	✓
s2 pre	✓	✓	✓	✓	✓	✓	✓	✓	✓	✓	✓
s2 post	✓	✓	✓	✓	✓	✓	✓	✓	✓	✓	✓
s3 post		✓	✓	✓	✓	✓		✓		✓	
all post	✓	✓	✓	✓	✓	✓	✓	✓	✓	✓	✓

Sets of entities tested:
global: GS13; 14 15; 19; 20; DE5; 4; 12; 13; 9a
g1: 100,106 (forms 1–3 and 11–13), 110, 116, 124 (form 0)
g2: 102, 104, 112, 124 (form 1), 126, 130
g3: 108, 118, 120, 122
g4: 114, 128, 140, 142, 144
a1: 212
a2: 202, 206, 210, 214, 216, 222, 106 (form 40)
a3: 220, 208, 218, 106 (forms 20 and 21), 228, 230, 106 (forms 31–38), 106 (form 63)
s1: 402 (forms 1, 7 and 16), 406, 308
s2: 412, 414
s3: 404, 410, 406 (form 15, 16 and 17)

Note: The NAVFAC tests were also conducted on Rucaps, Dragon, Artech (ECS) and GDS.

through this was the idea of an architecture for testing in Figure 10.3. This shows the use of a high-level format which embodies the content of IGES without its size and complex file structure. The architecture was intended to be applicable to other neutral formats by being table-driven, and was used successfully for similar software for VDA-FS 1.0. It addressed well the analysis of an IGES file, but necessarily not as clearly the method for testing preprocessor or postprocessor software, which is described in Chapter 10.

This testing underlined the need for tools and automatic test-file generation. In particular, a need for an IGES file comparator was highlighted. One of the first was developed in Germany [62].

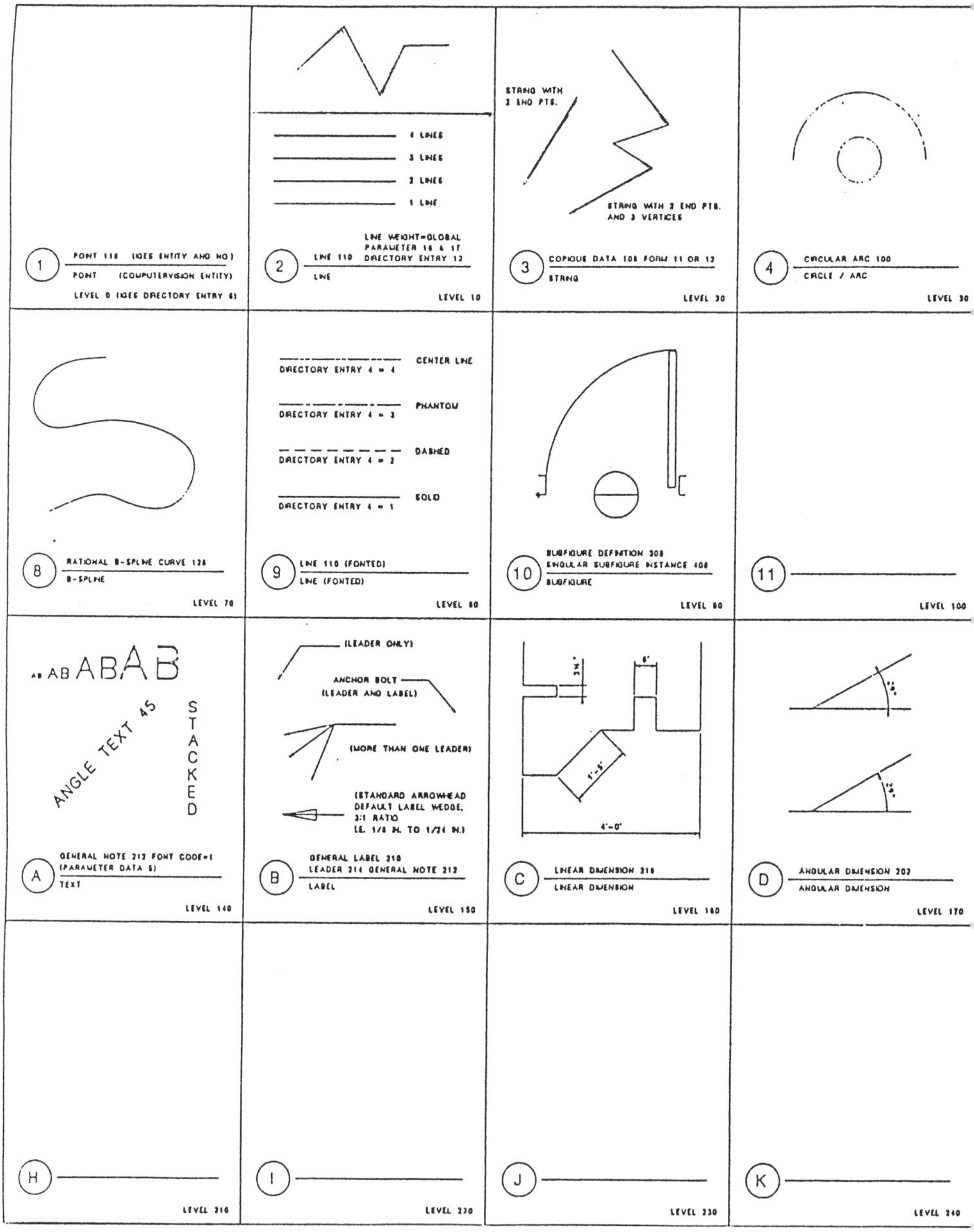

Figure 8.1 NAVFAC drawing.

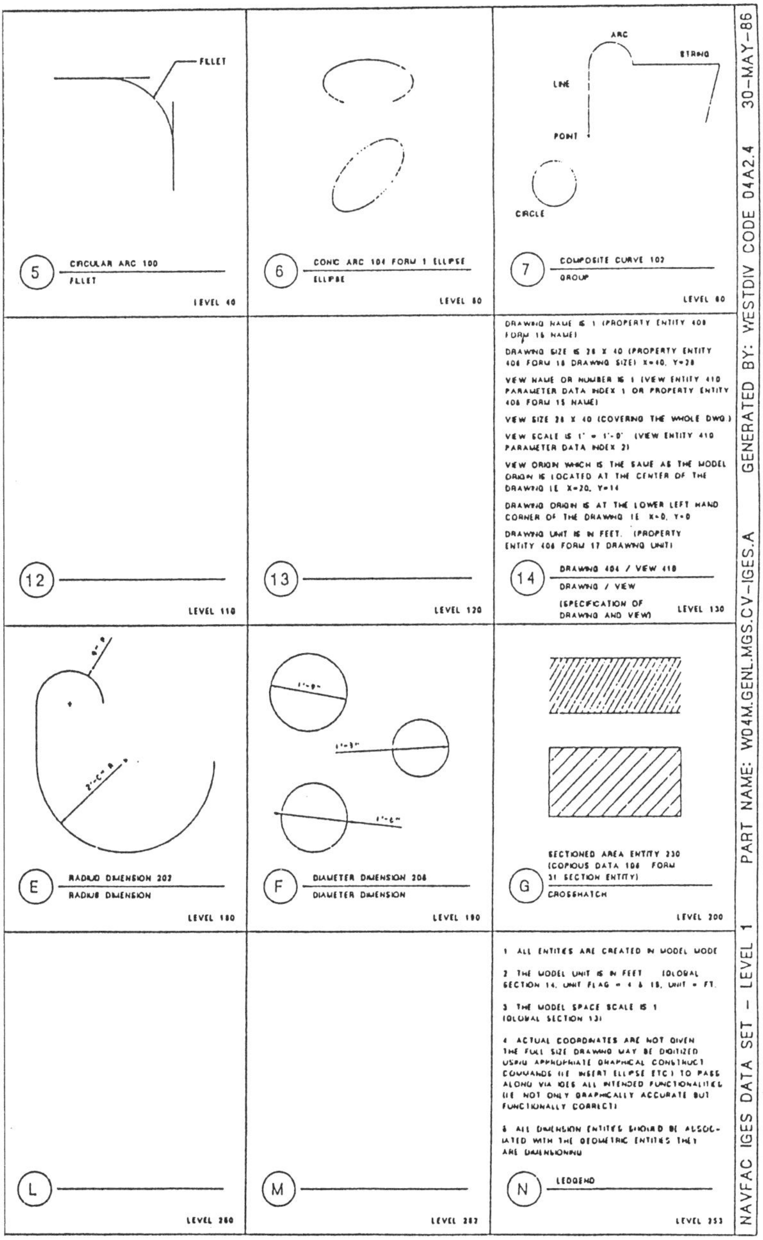

FLLET
5
CIRCULAR ARC 100
FLLET
LEVEL 40
6
CONIC ARC 104 FORM 1 ELLIPSE
ELLIPSE
LEVEL 50
ARC
STRING
LINE
POINT
CIRCLE
7
COMPOSITE CURVE 102
GROUP
LEVEL 60
12
LEVEL 110
13
LEVEL 120
DRAWING NAME IS 1 (PROPERTY ENTITY 408 FORM 16 NAME)
DRAWING SIZE IS 28 X 40 (PROPERTY ENTITY 406 FORM 16 DRAWING SIZE) X=40, Y=28
VIEW NAME OR NUMBER IS 1 (VIEW ENTITY 410 PARAMETER DATA INDEX 1 OR PROPERTY ENTITY 406 FORM 15 NAME)
VIEW SIZE 28 X 40 (COVERING THE WHOLE DWG.)
VIEW SCALE IS 1" = 1'-0" (VIEW ENTITY 410 PARAMETER DATA INDEX 2)
VIEW ORIGIN WHICH IS THE SAME AS THE MODEL ORIGIN IS LOCATED AT THE CENTER OF THE DRAWING I.E. X=20, Y=14
DRAWING ORIGIN IS AT THE LOWER LEFT HAND CORNER OF THE DRAWING I.E. X=0, Y=0
DRAWING UNIT IS IN FEET. (PROPERTY ENTITY 406 FORM 17 DRAWING UNIT)
14
DRAWING 404 / VIEW 410
DRAWING / VIEW
(SPECIFICATION OF DRAWING AND VIEW)
LEVEL 130
E
RADIUS DIMENSION 202
RADIUS DIMENSION
LEVEL 180
F
DIAMETER DIMENSION 206
DIAMETER DIMENSION
LEVEL 190
SECTIONED AREA ENTITY 230 (COPIOUS DATA 106 FORM 31 SECTION ENTITY)
G
CROSSHATCH
LEVEL 200
L
LEVEL 250
M
LEVEL 252
1 ALL ENTITIES ARE CREATED IN MODEL MODE
2 THE MODEL UNIT IS IN FEET (GLOBAL SECTION 14, UNIT FLAG = 4 & 15, UNIT = FT.
3 THE MODEL SPACE SCALE IS 1 (GLOBAL SECTION 13)
4 ACTUAL COORDINATES ARE NOT GIVEN THE FULL SIZE DRAWING MAY BE DIGITIZED USING APPROPRIATE GRAPHICAL CONSTRUCT COMMANDS (I.E. INSERT ELLIPSE ETC.) TO PASS ALONG VIA IGES ALL INTENDED FUNCTIONALITIES (I.E. NOT ONLY GRAPHICALLY ACCURATE BUT FUNCTIONALLY CORRECT)
5 ALL DIMENSION ENTITIES SHOULD BE ASSOCIATED WITH THE GEOMETRIC ENTITIES THEY ARE DIMENSIONING
N
LEDGEND
LEVEL 253
NAVFAC IGES DATA SET - LEVEL 1 PART NAME: WD4M.GENL.MGS.CV-IGES.A GENERATED BY: WESTDIV CODE 04A2.4 30-MAY-86

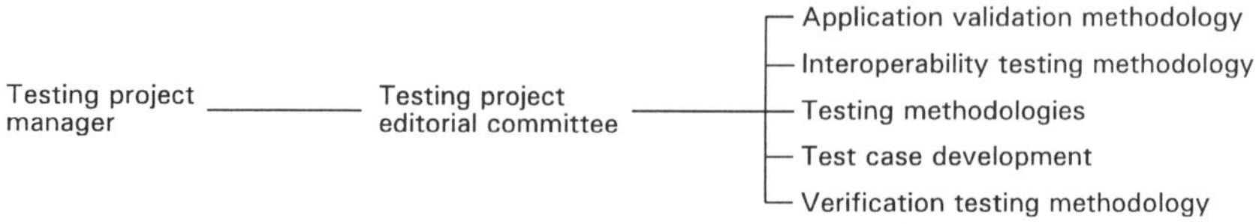

Figure 8.2 IGES/PDES Organization Testing Project.

8.5 IGES Committee and other US activities

In the late 1980s, the Testing Project of the IGES/PDES Organization (Fig. 8.2) was organized into committees to reflect the different types of testing (which are explained at the end of this chapter).

The Society of Automotive Engineers (SAE) also became involved, and helped to established a Verification Panel. This took the IGES test files and the methodology developed in the Testing Project, subcontracted universities to run the tests in conjunction with vendors, and reviewed the results. The findings from the first stage of tests confirmed that such an activity was not only manpower-intensive, but required those with a good understanding of the CAD system to be involved in the testing process. It also confirmed that software was needed in the production and analysis of test cases. The results helped to harmonize the activities in the USA and Europe. They also clarified the distinction between the conformance testing of an implementation and the peer review of a document; in the past, the two had been confused.

8.6 Advancing the application of IGES

The Society of Motor Manufacturers and Traders (SMMT) was a strong supporter of the NEDO initiative and also instrumental in establishing electronic data interchange (EDI) for orders and invoicing. Their interest came from increasing commercial pressures for manufacturers and suppliers to work more closely together in vehicle programmes.

In 1986, they invited all vendors with a presence in the United Kingdom to describe their IGES coverage and their plans to improve it. Their demand was that vendors should support all the entities in IGES. This initiative was well intentioned and the vendors co-operated but the demands were somewhat unrealistic [12].

Their next initiative was to extend the NEDO work to cover surface representation, an important requirement in automobile design and manufacture. This resulted in a report [120] which is still relevant today.

SMMT's major testing initiative is colloquially known as the Q-projects, short for quinpartite projects, because it consisted of sets of five

Table 8.2 Q-projects: sets of data and companies involved.

Pair of data-exchanging companies		Respective CAD systems	
Austin Rover Group	TRW CAM Gears	CV Cadds4X	IBM Cadam
Austin Rover Group	Hardy Spicer	CV Cadds4X	McDonnell Douglas Unigraphics
BR Engineering Ltd	Lucas Rail Products	Intergraph IGDS	IBM Cadam
Ford	Lucas Automotive	CV Cadds4X	IBM Cadam
Ford New Holland	Lucas Diesel Systems	Calma DDM	CV Cadds4X
Rolls-Royce	Westland Helicopters	IBM Cadam	CV Cadds4X

parties. These were two end-users, two CAD system vendors, and independent technical support. The objective was to create a method of working and testing which would enable swift and accurate data exchange with minimum technical input.

There were six sets and coincidentally six CAD systems involved, shown in Table 8.2. The work was supported by the Department of Trade and Industry and resulted in the report cited above. This included Figure 8.3 which is a methodology for testing during the planning stage of data exchange and indicates the value of communicating issues to the standards bodies (in this case the IGES committees). It is also essential to have the support of vendors: feedback to them is another feature of the method. The presence of an independent technical authority makes this feedback more acceptable, particularly as errors could be isolated and a small model produced which still contained the error, rather than a large file with the error buried somewhere within it.

The work is an example of interoperability testing, called "end-to-end" testing by the participants in the Q-projects. This is the users' prime interest, both in CAD data exchange and in network connections. The conclusion of the report [13] categorizes levels of testing and stresses the need for all levels.

8.7 VDA-FS testing

The first formal testing service was that offered by the CADCAM-Labor of the Kernforschungscentrum Karlsruhe (KfK) in Germany. This was a validation service for VDA-FS 1.0, the German surface exchange format. This was available in 1988 and defined a testing architecture for preprocessors and postprocessors which is shown in Figure 8.4. Not surprisingly, the focus was on the accuracy of the transfer of surfaces, because the representation conversions used may not be able to reproduce the data exactly. Deciding that two surfaces are the same is a non-trivial numerical problem. Test drawings were produced which can be represented

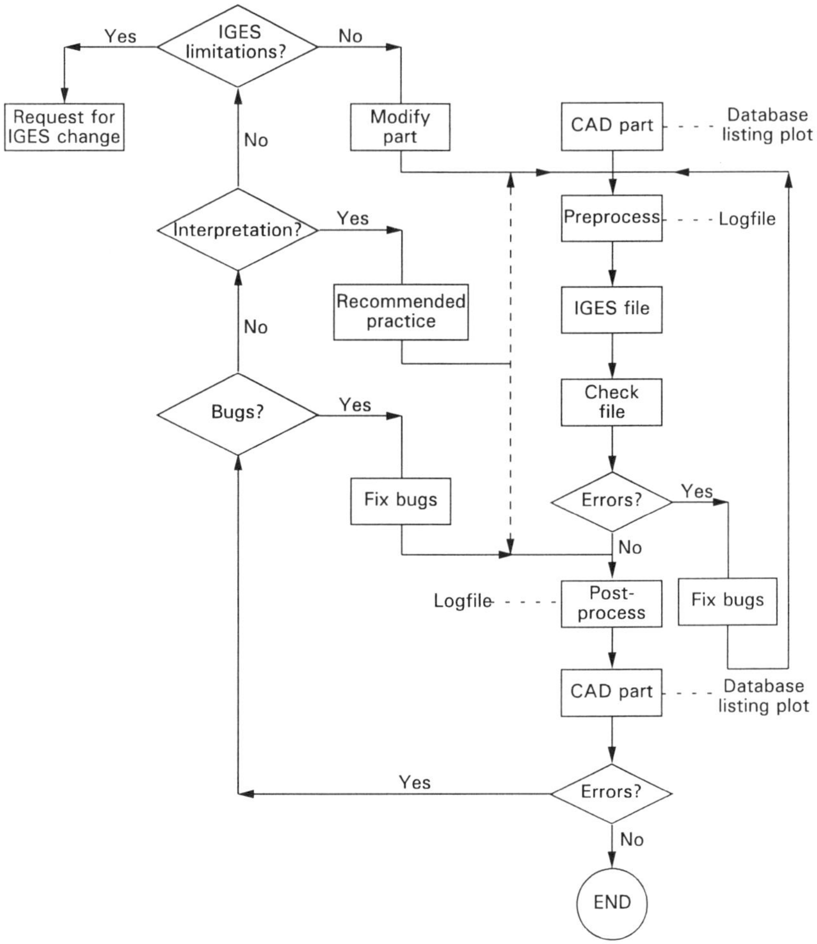

Figure 8.3 Q-projects: method for data transfer.

analytically to facilitate comparison. The service received accreditation from the appropriate German body (DEKITZ) in 1987. The procedures used in VDA-FS testing have been extended to form the basis of an IGES testing service.

8.8 Terminology

Clearly, a great deal of work by the user community has increased understanding of the problem of data exchange. Much has been fed back to the standards makers. The final section of this chapter is concerned with the education that will improve testing procedures. One lesson has been the

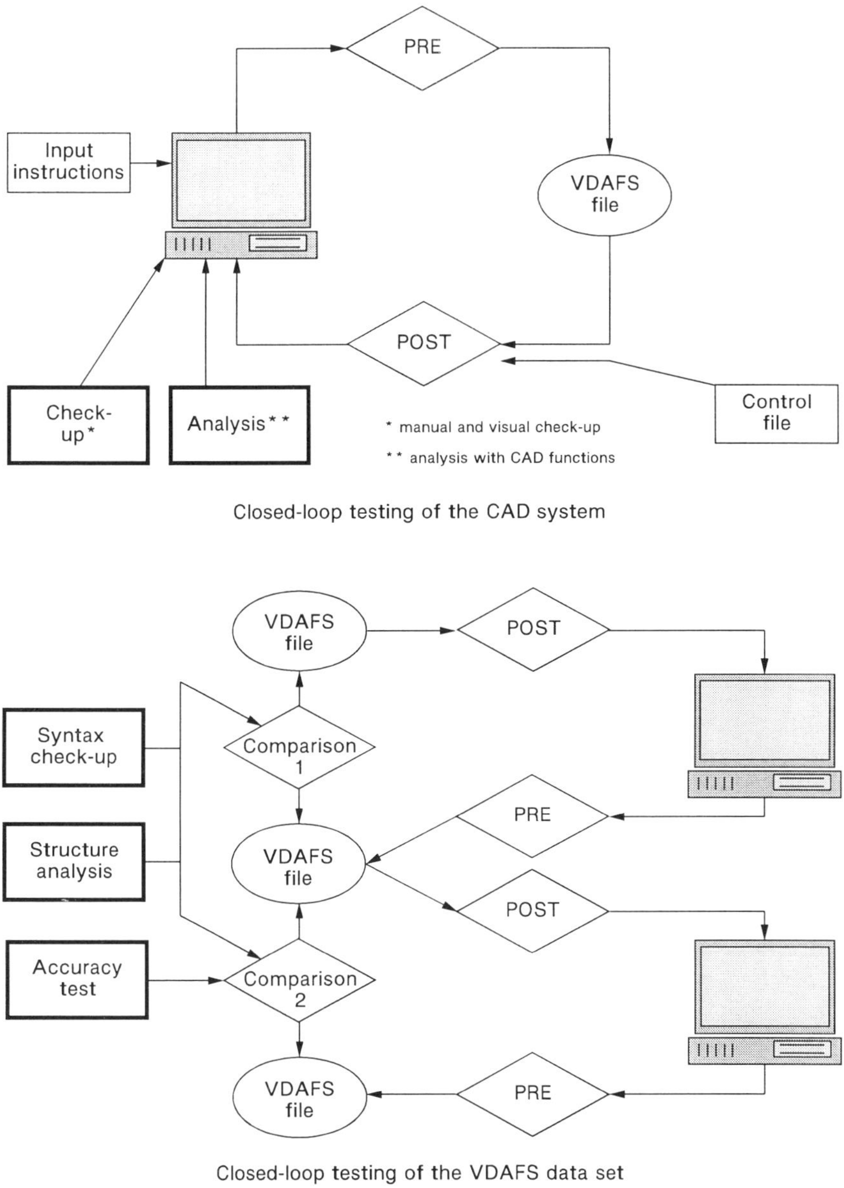

Figure 8.4 KfK testing method.

determination of a number of different testing levels and their relevance to reliable data exchange. First, we define testing terms that were used in an uneducated fashion throughout the 1980s and which led to unrealizable expectations in the user community. They also led to misunderstanding among those attempting to establish testing procedures.

8.8.1 Related terminology

Each term and its definition in the conformance-testing arena follow below.

Verification
This term is used to mean proving that an implementation is mathematically correct by taking a formal specification of the standard and successively transforming it, thus developing implementations. This process is closely related to techniques for program generation. For example, it is possible to use program transformations to derive line drawing algorithms which are functionally equivalent but more efficient than the original from which they have been derived [145]. For large systems this goal is a long way off. (Note that this definition conflicts with that of IEEE; see also "validation".)

Falsification
Falsification attempts to show that an implementation is incorrect. This cannot guarantee that it is correct but at least gives confidence that it has passed through some rigorous testing unscathed. This approach was already being used for GKS and the Ada programming language and is the basis for the conformance testing described in Chapter 10. (As Dijkstra has noted, program testing can be used to show the presence of bugs, but never to show their absence [36].)

Validation
Validation is testing performed by an independent authorized test centre. A reference system is needed with which to compare a candidate implementation. Although in common use during the 1980s, it is now used for the process of ensuring that the software tools used during conformance testing to check a candidate implementation are themselves correct. Consequently, it is used for activities within a test laboratory rather than with external clients, although the original definition, using a reference implementation, is still correct.

Certification
This is the action of legalizing the results of conformance testing by a body which is independent of the centre performing the testing, by issuing a certificate. This raises the issues of legal guarantees and liability, of the period of validity of a certificate, and of the availability of both reference results and test information. Planned testing will test preprocessors and postprocessors by a method based on falsification. There are currently no plans for issuing certificates. It is important that the process of conformance testing should itself be standardized.

Accreditation
This is the formalized initial and continuing process of ensuring a testing laboratory is competent to carry out specific (types of) tests. It covers the recognition of both the technical competence and the impartiality of a testing laboratory. Accreditation is normally awarded following successful laboratory assessment and is followed by appropriate surveillance.

8.8.2 Types of testing

Five levels are now recognized for software that translates between two data formats. This recognition has come from testing to make OSI products work, as well as from data exchange activities. The levels are conformance testing, user acceptance testing, interoperability testing, robustness testing and performance testing. Each is explained below.

Conformance testing
Conformance testing is testing a candidate product for the existence of specific characteristics required by a standard in order to determine the extent to which that product is a conforming implementation. Such testing focuses on the following requirements, which are described in Chapter 10:

(a) syntax;
(b) semantics;
(c) accuracy;
(d) completeness.

Conformance requirements are different for each entity and defined in a format-independent way.

User acceptance testing
User acceptance testing covers a broad spectrum of ways to test a processor, and requires the customer to define what is to be tested and what are the requirements. A user may require a processor to adhere to a user-defined subset of a neutral format. As with conformance testing, preprocessors and postprocessors are treated independently. User acceptance testing only really makes sense if conformance testing has already been performed on the processor.

Interoperability testing
Testing interoperability between two or more implementations means testing the complete transfer process, from the sending system to a neutral format and from the neutral format into the receiving system. Due to the very large number of ways to operate two systems involved in a transfer process, the testing criteria must be well defined. Again, interoperability can only proceed sensibly when the processors involved have previously been tested for conformance.

All new business relationships involving CAD data exchange should use this testing to prove the route, with the objectives of the data exchange being clearly stated at the outset so that appropriate testing is carried out.

Robustness testing
To ensure practical effectiveness, the conformance-tested postprocessors must be able to handle neutral-format files with incorrect entities, i.e. to neglect the entity and continue processing the file. The postprocessor should give an error message detailing entity type, problem and location in the file. Both preprocessor and postprocessors should also be tested with large physical models. The aim is to check if there is some upper limit in the native database (to be preprocessed) or the neutral files (to be postprocessed).

Performance testing
Testing performance means measuring the usage of computing resources by the implementation. This includes file size, memory usage, number of transactions, throughput and responsiveness under various conditions. The processing of large models may result in high use of certain computing resources.

These types of testing are applied to a software implementation of a standard or specification. A further type, proficiency testing, is applied to laboratories to ensure that they are technically competent. It is used by the accreditation body to determine laboratory testing performance by means of inter-laboratory test comparisons or by testing systems for which the results have already been determined.

Lack of robustness and poor performance were major contributors to the reception that IGES received. Processors failed to process entities they did not recognize and long processing times and large file sizes were common. Conformance testing is not concerned with these issues which the user finds important. However, unless implementations are first tested for conformance, further testing is not well founded. Although conformance testing is a prerequisite for all processors, interoperability testing for products and user acceptance testing are increasingly being considered with the same formality that conformance testing methods have, to date, demanded.

8.9 Conclusion

Early public testing of IGES was little more than a demonstration that the technology worked. This was valuable because it showed that, with discussions beforehand, the technology could be made to work. Informal testing programmes took place in several locations, providing useful feedback to users and vendors. They also resulted in changes being made to the existing specifications, and a better understanding of how to specify the domain of testing (Ch. 9) and the procedures and software for formal conformance testing (Ch. 10).

Chapter 9

Subsets and application protocols

A major problem with data exchange is the lack of overlap in the sets of entities that vendors support. Even when the CAD systems appear functionally the same, vendors may implement the entities they choose. This problem was foreseen by the designers of the Graphical Kernel System (GKS) and should perhaps have been foreseen by the designers of IGES, although the issue was addressed by Liewald and Kennicott [90]. As a result, after the introduction of IGES, subsets became an issue. A subset is a collection of entities which supports a given functionality; vendors with CAD systems of that functionality should support a subset both in the preprocessor and the postprocessor.

There are three subset initiatives worthy of note:

(a) that undertaken by the IGES committees after the introduction of IGES 3.0;

(b) the MIL-D-SPECs; and

(c) the VDA-IS IGES subset from the German car manufacturers.

Before discussing these it is worthwhile looking at the subsetting of GKS. This sets down two independent axes of functionality, input and output, with three possibilities on each, as shown in Table 6.1. There are thus nine levels of implementation, each of which specifies exactly what functionality is supported. The standard includes a list of functions ordered by the level in which they appear. Implementations of GKS that lie between two defined levels are invalid.

The formation of subsets for product data exchange is expanded in Figure 9.1 according to

(a) application area (discipline or industrial sector);

(b) product definition stage (of the life-cycle);

(c) geometric functionality.

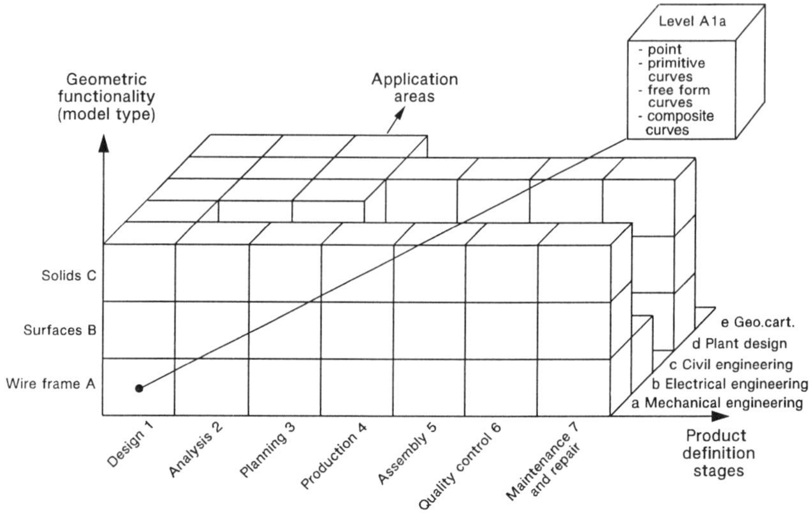

Figure 9.1 German proposal for STEP subsetting.

This had similar German origins to the GKS ideas. The formation of subsets of a neutral format would have three stages:

(1) definition of the subset axes and subdivisions, including subset names and functional description;
(2) development of an entity list without prejudging the association of entities into subsets;
(3) selection of entities for subsets determined by the practical requirements of the application area, product definition stage and geometric functionality.

These proposals were made in a position paper [149] to the Zürich meeting of ISO TC184/SC4/WG1 in 1985 as a basis for organizing STEP entities in the logical layer. Today it still looks at least as relevant to the subsetting of second-generation formats as to the STEP standard.

9.1 IGES proposals

In 1986, work started on the definition of application subsets for IGES.

IGES can support a number of applications; although it was originally developed for the transfer of engineering drawings and three-dimensional wireframes, there is now provision for electrical and finite-element modelling and for surface and solid representation.

The entities within IGES are (still) classified as geometry, annotation, definition or structure. A Request for Change was submitted which used this classification and divided each class of entities into IGES subsets. In addition, two application subsets were suggested, for finite-element

Table 9.1 The original proposal for IGES subsets (RFC 320A).

Set name	Contents
g1	100; 106 forms 1, 2, 3, 11, 12, 13; 110; 116; 124 form 0
g2	102; 104; 112; 124 form 1; 126; 130
g3	108; 118; 120; 122
g4	114; 128; 140; 142; 144
a1	212 fonts 1, 1001, 1002
a2	106 form 40; 202; 206; 210; 214; 216; 222
a3	106 form 20, 21, 31–38; 208; 218; 220; 228; 230
s1	308; 402 forms 1 and 7; 408
s2	412; 414
s3	404; 406 forms 15, 16 and 17; 410
s4	304; 310
s5	306; 600 and above
s6	406 form 12; 416
s7	320; 420
s8	302; 402 forms 5001 and above; 406 forms 5001 and above
FE	124 forms 10, 11 and 12; 134; 136; 138; 406 form 11; 418
electrical	102; 108 forms –1 and +1; 124; 125; 132; 212; 302; 308; 312; 402 form 18; 406; 408 *plus* subsets g1, s2 and s7

work and electrical schematics, as shown in Table 9.1. The global and directory entry fields were also addressed, with the lowest subset being mandatory for a processor to support.

The subsets were used for testing purposes, but were overtaken by the work on application protocols, in which the *use* of each entity is also described.

9.2 MIL-D specifications

The first widely advertised subsets were the MIL-D specifications of the US Department of Defense (DoD), which were originally published in 1987 [96]. A revised version appeared in 1992 [97]. The DoD is able to issue edicts and impose its requirements quickly whereas, as explained earlier, IGES has long-winded but robust procedures. The MIL-D-28000 Class I ("Technical publications") is shown in Table 9.2. Use of IGES for drawings in reports appears useful if IGES is the archival form. There are proponents of the simpler computer graphics metafile (CGM) in technical publications and it will probably be that both are used as appropriate, considering that they are both part of CALS [98].

MIL-D 28000 Class II ("Engineering drawings") is listed in Table 9.3. It collects the entities needed for both draughting and surface representation. This reflects some lack of clarity and compromise that STEP

Table 9.2 MIL-D-28000A: subset for technical publications.

Entity	Form	Description
0		null
100		circular arc
102		composite curve
104		conic arc
	0	determined by parameters
	1	elliptic
	2	hyperbolic
	3	parabolic
106		copious data
	11	2D linear path
	63	simple closed planar curve
110		line
112		parametric spline curve
124		transformation matrix
	0	orthonormal rotate/ translate
126		rational B-spline curve
	0	determined by parameters
	1	line
	2	circular arc
	3	elliptical arc
	4	parabolic arc
	5	hyperbolic arc
212		general note
	0	simple note (default)
	1	dual stack
	2	imbedded font change
	3	superscript
	4	subscript
	5	super-/subscript
	6	multiple stack/left justified
	7	multiple stack/centre justified
	8	multiple stack/right justified
	100	simple fraction
	101	dual stack fraction
	102	imbedded font change/ double fraction
	105	super-/subscript fraction
230		sectioned area
	0	standard cross-hatching
308		subfigure definition
404		drawing
	0	translate and scale
406		property
	15	name
	16	drawing size
	17	drawing units
	18	inter-character spacing
408		singular subfigure instance
410		view
	0	orthographic parallel
412		rectangular array subfigure instance
414		circular array subfigure instance

attempts to remove and the German VDA-IS avoids. In the "bottom-up" way that IGES subsets have been defined and because of the islands of automation that are the view of computerization that most engineers see, the relationship of these islands in a coherent framework is still exercising the STEP developers.

MIL-D-28000 Class III ("Electrical/electronic") is presented in Table 9.4 and that for Class IV ("Geometry for NC manufacturing") in Table 9.5.

The MIL-D specifications mandate only that no entities other than those shown in the lists will appear in a file from a processor that claims conformance. It does not mandate that all the entities are to be implemented. Thus, users who wish to exchange data have to use only the intersection of the entity sets, which is the lowest common denominator.

Table 9.3 MIL-D-28000A: subset for engineering drawings.

Entity	Form	Description
0		null
100		circular arc
102		composite curve
104		conic arc
	0	determined by parameters
	1	elliptic
	2	hyperbolic
	3	parabolic
106		copious data
	11	2D linear path
	12	3D linear path
	20	centreline through points
	21	centreline through circle centres
	31	section: solid
	32	section: solid pairs
	33	section: solid + dash
	34	section: solid quads
	35	section: solid + dash + solid
	36	section: dash
	37	section: solid + perpendicular solid
	38	section: solid + perpendicular dash
	40	witness line
	63	simple closed planar curve
108		plane
	0	unbounded
	1	bounded
110		line
112		parametric spline curve
114		parametric spline surface
116		point
118		ruled surface
	0	equal relative arc length
	1	equal relative parametric values
120		surface of revolution
122		tabulated cylinder
124		transformation matrix
	0	orthonormal rotate/translate
	1	orthonormal rotate–reflect/translate
126		rational B-spline curve
	0	determined by parameters
	1	line
	2	circular arc
	3	elliptical arc
	4	parabolic arc
	5	hyperbolic arc
128		rational B-spline surface
	0	determined by parameters
	1	plane
	2	right circular cylinder
	3	cone
	4	sphere
	5	torus
	6	surface of revolution
	7	tabulated cylinder
	8	ruled surface
	9	general quadric surface
130		offset curve
140		offset surface
142		curve on a parametric surface
144		trimmed parametric surface
202		angular dimension
206		diameter dimension
210		general label
212		general note
	0	simple note (default)
	1	dual stack
	2	imbedded font change
	3	superscript
	4	subscript
	5	super-/subscript
	6	multiple stack/left justified
	7	multiple stack/centre justified
	8	multiple stack/right justified
	100	simple fraction
	101	dual stack fraction
	102	imbedded font change/double fraction
	105	super-/subscript fraction
214		leader (arrow)
	1	wedge
	2	triangle
	3	filled triangle
	4	no arrowhead
	5	circle
	6	filled circle
	7	rectangle
	8	filled rectangle
	9	slash
	10	integral sign
	11	open triangle
216		linear dimension
	0	undetermined form
218		ordinate dimension
	0	witness line or leader
220		point dimension
222		radius dimension
	0	single leader format
	1	multiple leader format
228		general symbol
	0	general symbol
	1	datum feature
	2	datum target
	3	feature control frame
230		sectioned area
	0	standard cross-hatching
304		line font definition
	1	subfigure pointer
	2	repeating pattern
308		subfigure definition
314		colour definition
402		associativity instance
	3	views visible
	4	views visible, colour, line weight
	7	unordered group without backpointers
404		drawing
	0	translate and scale
406		property
	1	definition levels
	3	level function
	5	line widening
	15	name
	16	drawing size
	17	drawing units
408		singular subfigure instance
410		view
	0	orthographic parallel

Table 9.4 MIL-D-28000A: subset for electrical and electronic applications.

Entity	Form	Description
0		null
100		circular arc
102		composite curve
104		conic arc
	0	determined by parameters
	1	elliptic
	2	hyperbolic
	3	parabolic
106		copious data
	11	2D linear path
	12	3D linear path
	20	centreline through points
	21	centreline through circle centres
	40	witness line
	63	simple closed planar curve
108		plane
	-1	bounded void
	0	unbounded
	1	bounded
110		line
112		parametric spline curve
116		point
124		transformation matrix
	0	orthonormal rotate/translate
	1	orthonormal rotate–reflect/translate
125		flash
	0	defined by referenced entity
	1	circular
	2	rectangle
	3	doughnut
	4	canoe
126		rational B-spline curve
	0	determined by parameters
	1	line
	2	circular arc
	3	elliptical arc
	4	parabolic arc
	5	hyperbolic arc
132		connect point
202		angular dimension
206		diameter dimension
210		general label
212		general note
	0	simple note (default)
	1	dual stack
	2	imbedded font change
	3	superscript
	4	subscript
	5	super-/subscript
	6	multiple stack/left justified
	7	multiple stack/centre justified
	8	multiple stack/right justified
	100	simple fraction
	101	dual stack fraction
	102	imbedded font change/double fraction
	105	super-/subscript fraction
214		leader (arrow)
	1	wedge
	2	triangle
	3	filled triangle
	4	no arrowhead
	5	circle
	6	filled circle
	7	rectangle
	8	filled rectangle
	9	slash
	10	integral sign
	11	open triangle
216		linear dimension
	0	undetermined form
218		ordinate dimension
	0	witness line or leader
220		point dimension
222		radius dimension
	0	single leader format
	1	multiple leader format
228		general symbol
	0	general symbol
	1	datum feature
	2	datum target
	3	feature control frame
230		sectioned area
	0	standard cross-hatching
302		associativity definition
308		subfigure definition
312		text display template
	0	absolute co-ordinates
	1	incremental co-ordinates
320		network subfigure definition
322		attribute table definition
	0	one-to-many
	1	one-to-one
	2	one-to-one with text display template
402		associativity instance
	1	unordered group with backpointers
	7	unordered group without backpointers
	12	external reference file index
	18	flow
	5001..9999	implementor-defined (302 required)
404		drawing
	0	translate and scale
406		property
	1	definition levels
	2	region restriction
	3	level function
	5	line widening
	6	drilled hole
	12	external reference file list
	15	name
	16	drawing size
	17	drawing units
408		singular subfigure instance
410		view
	0	orthographic parallel
412		rectangular array subfigure instance
414		circular array subfigure instance
416		external reference
	0	one definition from library file
	1	full referenced library file
	2	external logical references
420		network subfigure instance
422		attribute tale instance
	0	single row
	1	multiple row

Table 9.5 MIL-D-28000A: subset for NC.

Entity	Form	Description
0		null
100		circular arc
102		composite arc
104		conic arc
	0	determined by parameters
	1	elliptic
	2	hyperbolic
	3	parabolic
106		copious data
	1	data pairs in a plane
	2	data triples in a plane
	11	2D linear path
	12	3D linear path
	20	centreline through points
	21	centreline through circle centre
	40	witness line
	63	simple closed planar curve
108		plane
	-1	bounded void
	0	unbounded
	1	bounded
110		line
116		point
118		ruled surface
	1	equal relative parametric values
120		surface of revolution
122		tabulated cylinder
124		transformation matrix
	0	orthonormal rotate/translate
	1	orthonormal rotate–reflect/ translate
126		rational B-spline curve
	0	determined by parameters
	1	line
	2	circular arc
	3	elliptical arc
	4	parabolic arc
	5	hyperbolic arc
128		rational B-spline surface
	0	determined by parameters
	1	plane
	2	right circular cylinder
	3	cone
	4	sphere
	5	torus
	6	surface of revolution
	7	tabulated cylinder
	8	ruled surface
	9	general quadric surface
130		offset curve
140		offset surface
142		curve on a parametric surface
144		trimmed parametric surface
202		angular dimension
206		diameter dimension
210		general label
212		general note
	0	simple note (default)
	1	dual stack
	2	imbedded font change
	3	superscript
	4	subscript
	5	super-/subscript
	6	multiple stack/left justified
	7	multiple stack/centre justified
	8	multiple stack/right justified
	100	simple fraction
	101	dual stack fraction
	102	imbedded font change/double fraction
	105	super-/subscript fraction
214		leader (arrow)
	1	wedge
	2	triangle
	3	filled triangle
	11	open triangle
216		linear dimension
	0	undetermined form
218		ordinate dimension
	0	witness line or leader
220		point dimension
222		radius dimension
	0	single leader format
	1	multiple leader format
228		general symbol
	0	general symbol
	1	datum feature
	2	datum target
	3	feature control frame
304		line font definition
312		text display template
402		associativity instance
	3	views visible
	4	views visible, colour, line weight
	7	unordered group without backpointers
	9	single-parent
404		drawing
	0	translate and scale
406		property
	15	name
	16	drawing size
	17	drawing units
410		view
	0	orthographic parallel

Apart from the fact that there are as yet no independent bodies or accredited procedures to ensure that processors conform, the requirements only restrict vendor choice. They do not address or solve the real problem: that of "completeness".

In fact, even this first restriction is moderated because the specification allows "volunteer" entities "as long as these entity types are valid Y14.26M entities, are not necessary to the product data representation, and are solely for the purpose of regenerating the same development environment when the file is transferred back to the same CAD system which originally generated the file".

9.3 VDA-IS

VDA-IS was originally published in 1987 [154] and a revised version appeared in two parts in 1989 [155]. It defines seven subsets for the IGES entities as shown in Figure 9.2. There are three simple geometry subsets, two for annotation and structure and two application subsets which cover free-form geometry. A processor chooses to support at least one geometry and one other subset. The three classifications are hierarchical and a processor that supports G2 necessarily also supports G1. The geometry subsets (G1, G2, G3) support that geometry displayed in engineering drawings. The annotation subsets (B1, B2) supply the annotation and collection into drawings and views. The application subsets (AF1, AF2) support surface representation, in polynomial form and rational B-spline form respectively. This subsetting reflects more accurately and clearly the current functionality of computer-aided draughting and surface representation systems than the MIL-D specifications. More importantly, for a processor to claim support for given subsets, it must support *all* the entities in those subsets. This is the completeness requirement that is now recognized to be crucial.

Furthermore, prescriptions are included in VDA-IS for the conversion of entities so that the receiving postprocessor can "map" information, admittedly with loss or degeneration of content, into the lower subset that it supports. However, no guidance was given on what application requirement each entity is intended to fulfil. A polyline may still be used in a single given application for many and varied purposes (e.g. to represent both a curve and a character). These will appear indistinguishable to the receiving postprocessor. An important part of the requirement for application protocols, discussed later in this chapter, arose from the need to control this problem to maximize information content and also to facilitate testing for conformance.

Another important initiative associated with VDA-IS was the establishment of a project to test conformity to the VDA-IS specification. VDA

Geometry

G3

Type	Form	Name
106	3	Copious data (Coord, sextuple)
106	13	Copious data (Coord, sextuple vertices)
108	1,0	Plane

G2

Type	Form	Name
106	2	Copious data (coord, triple)
106	12	Copious data (coord, triple vertices)

G1

Type	Form	Name
100		Circular arc
102		Composite curve
104	0.1	Conic arc (unspecified, ellipse)
104	2.3	Conic arc (hyperbel, parabel)
106	1	Copious data (coord, pair)
106	11	Copious data (Coord, pair vertices)
110		Line
112	0	Parametric spline curve
116		Point
124		Transformation matrix
402	7	Group without backpointer

Annotation

B2

Type	Form	Name
108	0	Unbounded plane
402	3	View visible list
404		Drawing
406	16	Drawing size
410		View

B1

Type	Form	Name
106	31	Copious data (section)
106	40	Copious data (witness line)
202		Angular dimension
206		Diameter dimension
210		General label
212		General note
214	1,4	Leader (wedge, no arrowhead)
214	9,10	Leader (slash, integral sign)
216		Linear dimension
218		Ordinate dimension
220		Point dimension
222		Radius dimension

Application*

AF2* *Free-form geometry (rational)*

Type	Form	Name
126		Rational B-spline curve (rational form)
128		Rational B-spline surface (rational form)
302		Associativity definition
402	5189	Associativity instance/VDATOP

AF1* *Free-form geometry (polynomial)*

Type	Form	Name
126		Rational B-spline curve (polynomial form)
128		Rational B-spline surface (polynomial form)
142		Curve on parametric surface
144		Trimmed parametric surface

Figure 9.2 VDA-IS subsets.

produced test guidelines which, taken in conjunction with the specifications, result in all options having been carefully specified in each particular application domain. This pioneering work is not yet available for other second-generation formats and is an antecedent of conformance clauses in STEP application protocols.

9.4 Context dependence

Examples of the problems in transfer of a particular drawing are discussed fully in Chapter 3. Figure 3.10 appears at first inspection of the drawing to have been successful. (It must be borne in mind that identical drawings do not mean that all the information, particularly structure such as view relationships and higher-dimensionality geometry, has been transferred, as shown in Figure 3.11.) Closer scrutiny reveals that some text has been changed from lower-case to upper-case. One of the systems, or perhaps the magnetic-tape utility that transfers from disk to tape (or *vice versa*), may have caused the change.

Whether this matters or not depends on the context. In some cases it may be acceptable; for example, if the requirements are that "the text is readable", "the position of the text does not change", or perhaps "the text box does not change in size". These may apply for engineering drawings where the requirement is readability which may be impaired if text obscures the drawing. However, for text publishing it may be vital to have the text transferred precisely, including the exact font.

Most product data exchange specifications, including STEP, have a large number of general-purpose or context-independent entities which can be used in a variety of applications. This provides a richness which makes the specification useful in a wide range of applications but is at the root of many current problems.

As an example, consider the line entity. It may be used to capture a genuine two-dimensional line or a segment of a higher-order curve or a character in a technical illustration. It may be used to represent an edge in a wireframe drawing, a pipe in a piping and instrumentation diagram, an electrical connection, or an axis of symmetry important to a machining application. A receiving postprocessor may need to distinguish between all of these and to convert the information into its own more relevant form. What is needed with each entity is additional specification which adds constraints for use in a particular application domain which is itself well defined and constrained. Context dependence is an important part of application protocol methodology which will be described next.

9.5 Application protocols

Application protocols give not only a list of entities but also the context in which they are to be used and a mapping indicating which particular task they perform in an application. (The term "application protocol" was first used by Liewald and Kennicott [90], but does not seem to reappear until work on IGES testing several years later.) The entities are the same general-purpose, context-independent entities that form the basis for the neutral format specification but they are now given a context and additional context-dependent constraints.

The idea of application protocols is now enshrined in the STEP architecture and methodology described in Part IV. Its origins in CAD data exchange have been indicated above. Concurrently, the open systems interconnection (OSI) testing work [84, 85] was developing methods using application profiles. These have the same purpose, in trying to specify conformance criteria for functional subsets of standards and specifications.

The value of the idea was accepted by the IGES committees which are now committed to producing application protocols. The Application Validation Methodology Committee has developed a guideline document for the developers of application protocols [112]. The National Institute of Standards and Technology has coordinated a project to produce an application protocol of interest to the AEC community. This supersedes their work on subsets which has been discontinued.

9.6 Conclusion

The early second-generation specifications did not indicate which entities should be supported for a particular application. As a result, vendors made an *ad hoc* choice, which resulted in a mismatch of coverage between processors when an exchange took place. VDA-IS introduced subsets of IGES entities, requiring vendors to implement all entities within a particular level, and providing conversion prescriptions to entities in a lower level. The US military specifications also defined subsets for IGES. Later, application protocols documented the *use* of each entity – that is, the requirement which it fulfilled. This work was used in the development of STEP standard for exchange of product data.

Chapter 10

Conformance testing

Conformance testing is defined as "the testing of a candidate product for the existence of specific characteristics required by a standard in order to determine the extent to which that product is a conforming implementation" [146]. Conformance testing is not new and in information technology has been applied to language compilers and networking products. The European Community's IT Directorate (DG XIII) has supported several conformance-testing initiatives. The CTS1 and CTS2 programmes have developed methods and established test services for a number of standards, including telecommunications and graphics, both European and international [31].

10.1 Conformance testing of CAD data-exchange processors

In the context of product data exchange, or more accurately the second-generation CAD data-exchange processors, conformance testing means that a candidate product must

(a) conform to the syntax specified;
(b) preserve the semantics or meaning that this is intended to convey; and
(c) confirm that the meaning is accurately and completely conveyed.

In the case of VDA-IS, the completeness requirement and prescriptions for converting from one entity to another must also be checked.

Conformance testing is a prerequisite for all other, more user-oriented, testing. Most particularly, user acceptance testing is conformance testing plus robustness, interoperability and performance testing (see Ch. 8) as defined by the user.

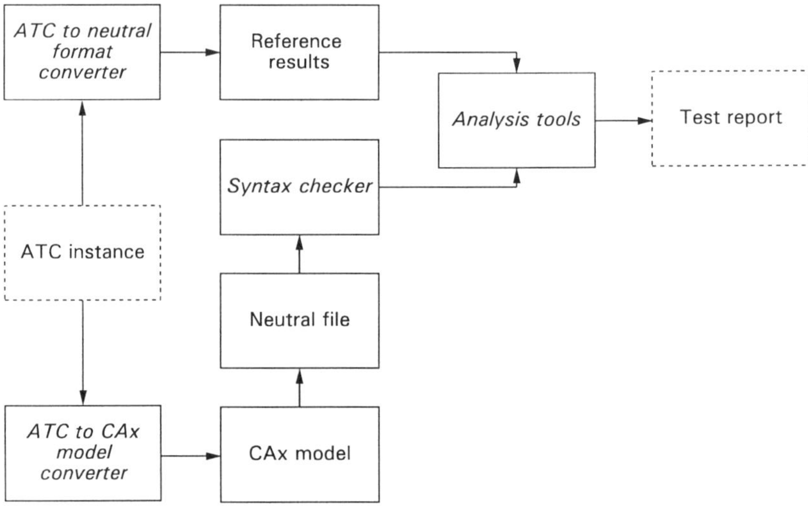

Figure 10.1 Preprocessor conformance testing (physical file).

In order to perform conformance testing, one thus has a candidate product, called the implementation under test (IUT), and the specification that comprises the standard. For CAD data-exchange processors, the preprocessor and the postprocessor are two independent IUTs. Test procedures, including test data, must be documented. These abstract test cases (ATCs) exercise the IUT and demonstrate

(a) that a preprocessor produces files that are as described by the standard, from input to the CAD system that has the same functionality (see Fig. 10.1); or
(b) that a postprocessor reads a file conforming to the standard and reproduces the information accurately and completely in its own CAD system (see Fig. 10.2).

These tests must address the following categories:

- syntax: for the preprocessor, check that the file format is as prescribed, which includes the number of fields, the data types in those fields, and layout where this is important; for the postprocessor, there is no syntax to check because the observed outcome is a model in the CAx system;
- semantics: this is more concerned with the contents of the fields such as sizes (Is the radius non-zero?) and the existence of relationships (Is an entity in a composite curve of an appropriate type? Does a group entity point to more than one other entity? Has a circle or some other degenerate replacement been received?);
- accuracy: this is checking that particularly a curve or surface can generate the same data points as the original, or that the connectivity relationship between elements in a schematic is the same as that in the original.

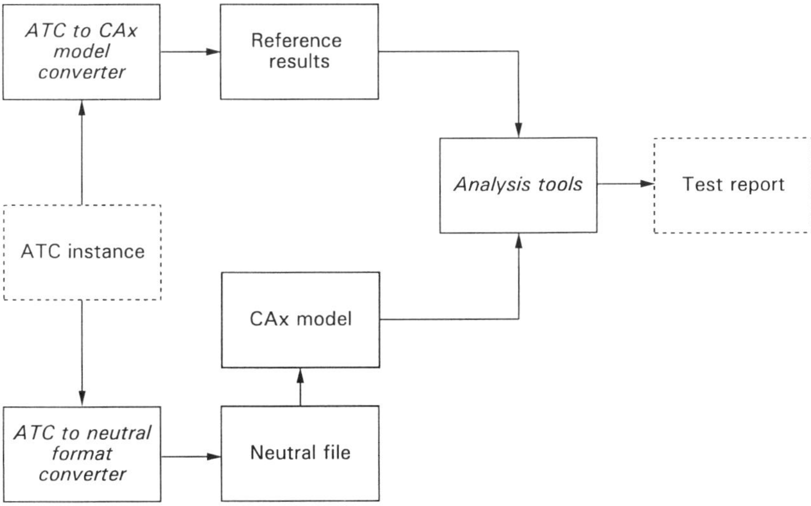

Figure 10.2 Postprocessor conformance testing (physical file).

Clearly, the behaviour of the CAD system is an integral part of what is being tested, both for preprocessor and postprocessor. Also, there is some human interpretation in inputting the test cases for preprocessor testing and there is a need to access and interpret the CAD system database for postprocessor testing. Therefore, interaction with the processor vendor is essential.

In order to test that a processor conforms to a particular standard a large number of test cases are required to reflect the number of aspects that need testing. The issue of whether each aspect needs a separate test case or how these should be combined is still contentious. The formulation and generation of test cases is discussed below. Conformance testing includes giving help to vendors to ensure that their processor performs adequately. The aim is to provide test cases to try on-site so the vendor only presents a processor for testing when convinced that it will pass. Thus one needs to be able to regenerate instances of a particular test with only the data values changed. This leads us to the idea of abstract and executable test cases.

The number of test cases demands software to generate them automatically; experience with IGES has already demonstrated this. There are other parts of testing that are enhanced by some automation, including syntax, semantic and accuracy checking for the preprocessor. Further, the generation of instructions for prescribing input to the CAD system and templates for evaluating the information that has been deposited by the postprocessor into the CAD system may also be automated in part. This list is not exhaustive, and the architecture of a testing method and the software that it demands are presented in Section 10.9.

For a testing service to have value both for the vendors and end-users it must be demonstrably independent and its reports must be widely recognized and credible. The provision of such an environment including accreditation and harmonization is the subject of Section 10.7.

The result of conformance testing is a report which the client[3] or some other person submitting the processor for testing may use as he wishes (subject to the contract between the client and the test laboratory). Currently, it is not planned to provide certification for processors, which would have legal implications. Current specifications and testing methods are not sufficiently robust to instil enough confidence for testing laboratories to be willing to consider taking on that additional legal responsibility.

10.2 Conformance to what?

The requirements for user acceptance are necessarily set by the user. They may include requirements that the software being tested can interoperate with another particular CAD system or set of systems, possibly for a given set of data. They may also include performance and robustness tests. These sorts of tests are invariably different for different users; consequently, they cannot be standardized.

On the other hand, requirements tested during conformance testing are set down in the standard itself. This means that anything not specified in the standard cannot legitimately be considered as part of conformance testing. A vendor should support what is stated in the standard; conformance testing checks that this is so. Once a processor has undergone conformance testing, this can be used as the basis for further, user-specified, testing.

10.3 Ambiguities

As described in the previous chapter, only VDA has attacked both conformance requirements and completeness in testing would-be conforming processors. Elsewhere, the problems of ambiguity and differences of interpretation remain.

Most existing specifications are written in a natural language such as English. In IGES, phrases such as "any parametric curve", "any curve that can be parameterized", and "any curve which is two-dimensional" are prevalent. There is no definition of the set of parametric curves, so the choice is arbitrary; consequently, it is usually determined by personal preference or experience. Are the first two the same? Should the last be restricted to truly "two-dimensional" entities or are those three-dimensional ones acceptable which happen to be planar? Does it have to have a z-coordinate set to zero, a constant z-plane, or is any plane acceptable? In IGES there are examples of each.

10.4 Options

Even within a well defined standard, options may exist deliberately. These have several forms, but give the implementor some flexibility in what is implemented, and necessarily impact both on what can be transferred and what needs to be tested.

The options within a composite curve in IGES 4.0 allow it to be composed of any or all of the following entities:

(a) circular arc;
(b) conic arc;
(c) line;
(d) parametric spline curve;
(e) point;
(f) rational B-spline curve;
(g) connect point.

(IGES 5.0 extended this to include any parameterized curve entity, which introduces copious data and offset curve entities.)

A vendor may choose to implement any or all of these. One can understand that he may not implement any option of which there is no equivalent in his system. More subtly, although he may support a particular entity in his CAD system, he may not allow it to be a constituent in the equivalent of the composite curve. Any test of this system must know this and only test the composite curve with those entities that can be included in each particular system.

Another case that produces options is the defaulting mechanism. The point dimension in IGES may have its text underlined (which is the default) or enclosed by a circle or a hexagon. There are therefore three options which the vendor may choose to implement.

10.5 Equivalence between IGES and the CAD system

Conformance testing of IGES preprocessors requires a neutral format file to be generated from the entities constructed by the appropriate input to the CAD system. Conformance testing of IGES postprocessors requires examination of the database of the CAD system, which has been populated by reading a neutral-format file. Because there is no context for each entity, the instructions to the client are “do what it takes to create an IGES line entity” and “read this IGES line entity and indicate the resulting CAD construct". In most cases there will be a CAD system line, but not necessarily. Conformance testing cannot pronounce on whether the IGES line was created from something that was not obviously equivalent or that an IGES line generated something unexpected. This is because there is no requirement in the specification that conveys the *idea* that the IGES line should support. There are no rules for the mapping from IGES to CAD or CAD to IGES. There are many options, as suggested earlier.

To develop a testing service, criteria on which the testing is based have to be developed. The IGES/PDES Organization, particularly the Testing Project, has helped with this. Secondly, interaction with the client submitting a processor is essential. This is not unique. OSI (Ch. 6) has the concept of a PICS (Protocol Implementation Conformance Statement) and a PIXIT (Protocol Implementation eXtra Information for Testing). The PICS is a statement made by a client stating the capability and options that have been implemented in the IUT. The PIXIT is a statement that contains or references all the information (in addition to that given in the PICS) relating to the IUT and its corresponding system under test (SUT) which will enable the test laboratory to run an appropriate test suite against that IUT. Even though completeness may be required in principle by a standard, there may still be options or alternatives within it which will have to be enumerated as being supported or not by the implementation.

The information required from the client is, in addition to administrative details such as software and hardware details and contact names, the following:

(a) a list of all the possible entities and options which are available in the chosen version of the neutral format;
(b) restrictions on usage (e.g. only two-dimensional mode) or parameter ranges (e.g. no larger than 10);
(c) mappings between CAD system features and neutral format features and information on access to the CAD system data;
(d) information to allow running of the CAD system during testing.

The testing method developed by the CTS2 CAD data exchange project is described in Section 10.9. Such techniques were built into STEP from its inception.

10.6 Architecture and software tools

Because of the number of test cases that a typical CAD data exchange standard requires, many of the tasks undertaken during testing are repetitive. This can be ameliorated by using as much automation as feasible in any testing method. Use of automation enables a method and

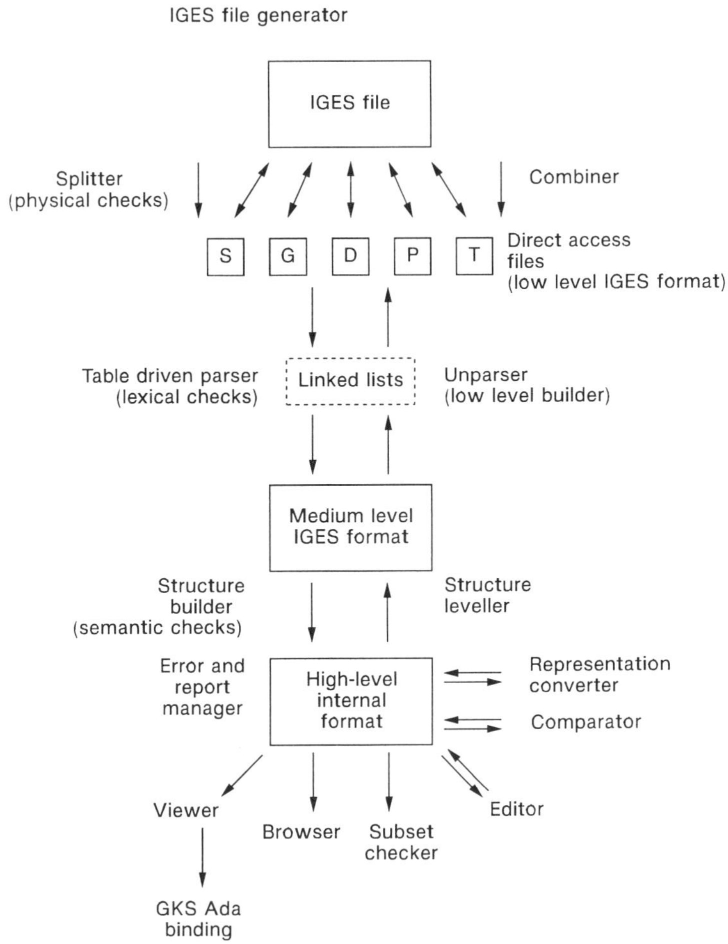

Figure 10.3 CADDETC software architecture.

procedures to be developed that can be adapted to test other CAD data-neutral formats with minimal changes. Properly designed software with that requirement can facilitate this and may be of value when the need for other similar conformance-testing services, for example SGML and CGM, is met.

The software that is required for conformance testing has evolved from that used for the informal testing carried out by users and technical support groups. An architecture proposed in that period is shown in Figure 10.3. This was used for IGES file analysis and started from receipt of an IGES file. By using a table-driven parser and a high-level format (HLF, also known as the "working form"), the architecture can be used for other similar neutral formats. The ideas have since been extended and used for a series of STEP tools. The idea of a high-level format which captures the essential information in a neutral-format physical file is part of the more formal conformance-testing methodology and of many independent commercial STEP-based software products: the PDDI project [113] used a similar approach. At one stage it had been hoped that this working form would be STEP itself, but conceptual differences between second-generation formats and STEP ruled against this. However, wherever possible (e.g. for geometry), STEP is now used as the reference form.

We first describe the software components in Figure 10.3:

- creation of working form: reads in the IGES file and stores the contents in active data structures;
- syntax checking: during the reading of the IGES file, checks that the constructs conform to the syntax and semantics given in the specification;
- comparison: compares the reference results with the observed outcome;
- representation conversion: converts from one form of a construct to another, but with possible loss of information (this is also known as flavouring);
- viewing: presents the constructs graphically (a first coarse check);
- editing: changes individual fields of the constructs, particularly those that have errors or are known to cause problems to a receiving system (this would call the representation converter for changes which involved an entity substitution); it would not be used in conformance testing.

Figures 10.1 and 10.2 show how two CAD systems (one with a preprocessor, CAD to IGES, and one with a postprocessor, IGES to CAD) undergo conformance testing. In testing these two are independent and treated separately. Some vendors only produce one of the pair. Also, good testing practice does not allow use of an unproven piece of software to facilitate testing of another piece. It is difficult to obtain the mappings between the CAD system and the neutral format, which are crucial to testing current standards. Consequently, it has been suggested that testing be based on a reflection test, such as that performed in the VDA-FS and VDA-IS testing method, shown in Figure 8.4 and developed by KfK CADCAM-Labor. This is of particular interest to users, not least because it has often been found not to work, contrary to one's initial expectations. However, it does not provide useful information about the conformance of a processor even if information returns to the sender as sent: symmetric errors can occur in a pair of processors, and if an error is identified, it cannot be isolated in one particular processor.

In order to test that a postprocessor conforms to a neutral format, files in that format must be generated to test every aspect of that format. Details of what is an aspect, commonly called a test purpose, and how test cases are defined, are described in the section on abstract test cases below. The files are then read, one by one, into the postprocessor under test. The major problem is how to access and evaluate the contents of the CAD database that results. Liaison with the client before testing must establish the CAD entities generated from a given neutral-format entity so that the form in which the information will be presented is known. There may be one of several "acceptable" forms (or templates) of the information, the preferred one of which is that documented in the specification.

Software is available which enables the automatic generation of executable test cases, representation conversion, and comparison between the reference results and the observed outcome. For the preprocessor, the human operator is instructed how to use the CAD system to generate entities that when preprocessed will produce the desired neutral-format file (i.e. containing the entity forms being tested). This file is checked for syntax and converted into the working form; this form of the observed outcome is compared, by software, with the reference results. For the postprocessor, an IGES file is read into the system, populating the database. The user is instructed to interrogate the model to elicit specific information which can then be checked against the values in the reference results. It should be noted that the template form used to provide input for the preprocessor test is identical to that populated by the postprocessor.

10.7 Requirements of national accreditation

Conformance testing may be performed by anyone. It is a requirement of the European-funded CTS projects (which develop conformance-testing methodologies and establish testing services) that the procedures, methods and test cases be made available at reasonable prices to anyone. It is envisaged that large user companies may find all or part of the packages worth investment. Even though the methodology allows for prevalidation by clients, vendors may find the test suite helpful during the actual development of processors.

It is possible that many testing laboratories may offer services although the cost of developing the methodologies and running services compared with the price clients may be willing to pay makes the chances of large profits unlikely. In order to ensure that testing laboratories are seen to be independent and that their reports are correct and widely recognized as valuable, a worldwide system of accreditation of testing laboratories and mutual recognition of reports has been and is being established. The ECITC guide [46] lists the accreditation bodies in a selection of countries and the status of recognition agreements. Laboratories that wish to provide an accredited testing service apply to their national accreditation body and then have to follow a series of

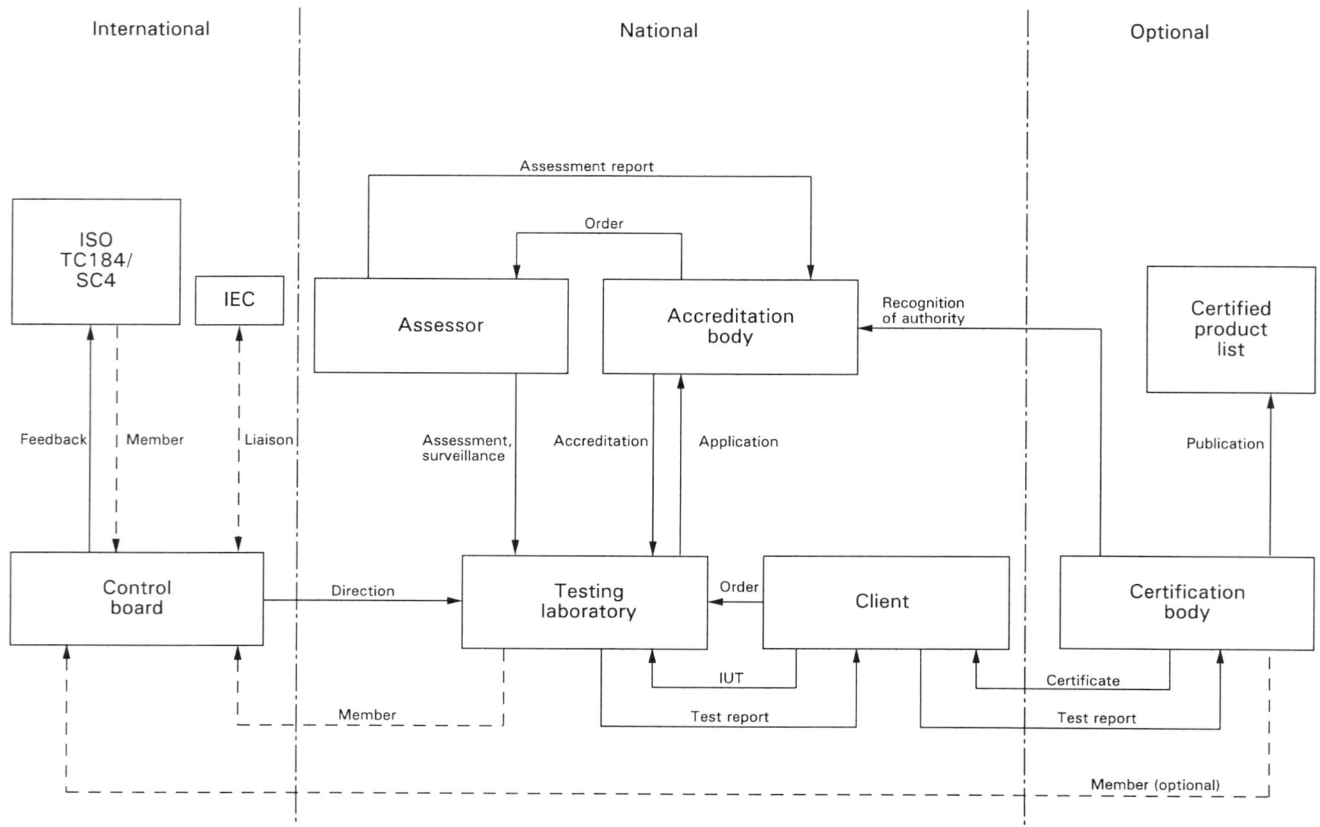

Figure 10.4 Infrastructure for conformance testing.

procedures culminating in accreditation for a particular testing service. It was envisaged that all countries would operate the same, or at least equivalent, procedures for accreditation, but regrettably this has not happened. The objectives of the procedures are

(a) to ensure that the laboratory is independent;
(b) to ensure that its working practices adhere to national [23], European [49] or international [83] quality standards;
(c) that the proposed testing method be technically and procedurally sound and will achieve its aims; and
(d) that in all ways (e.g. size, stability) the laboratory be fit to operate.

The first time, gaining accreditation can take between six months and two years from initial application if there are no delays due to the laboratory. The accreditation body will make regular rechecks of the laboratory to maintain its rigour.

Accreditation is the first step towards mutual recognition internationally. Clients, which (as noted earlier) are usually vendors, will expect to have to pay for testing only once for each product and have that test report accepted wherever that product is marketed. There are examples of mutual recognition shown in the ECITC guide [46,] which covers information technology testing and certification. However, many of the procedures required for accreditation and mutual recognition are common to any form of testing, and are not limited to software testing [83].

The European Union is working hard to establish mutual recognition within all member states and EFTA countries. To facilitate this, the CTS programmes are generating harmonized test services by insisting that one of the objectives of each project be that two laboratories set up services using the same methodology and test suite, to be as harmonized as possible. More recently, the accredited IGES conformance-testing service developed at CADDETC has been licensed in the USA by the National Institute of Standards and Testing (NIST).

The infrastructure in which conformance testing takes place is depicted in Figure 10.4.

10.8 The abstract test case

The concept of abstract test cases (ATC) and the methods emerging to support this approach are fully described in technical reports [32, 33] of the CTS2 CADCAM project for the second-generation formats, and later for STEP [146].

Abstract test cases are written in a formal definition language and contain

(a) test purpose;
(b) test case identifier;

```
UNIQUE e100x1

REFERENCE IGES 4.0 section 3.2 (pp72-73); p25

CRITERIA IGES file shall have no syntax errors
CRITERIA entity 100 (circular arc) shall be present
CRITERIA all parameter data fields shall be within tolerance

c := circular_arc 0.0, 5.0, 5.0, 35.0, 5.0, 10.85271, 34.42356
```

Figure 10.5 Extract from sample abstract test case.

```
                                                                        S      1
1H,,1H;, 12HExample test,   6HZ.IGES,   5HLuigi,   11Hversion 4.0,      G      1
 32,  38,   6, 308,  15,  12HExample test,  1.00000E+00,   2,   2HMM,   G      2
    1,  0.00000E+00,  13H940126.090223,  1.00000E-06,  0.00000E+00,     G      3
  8HJon Owen,  37HCADDETC, University of Leeds, England,   6,   0;      G      4
    100        1        0        1        0        0        0        000000000D      1
    100        0        0        3        0                 circ arc     1D      2
    100,  0.00000E+00,  5.00000E+00,  5.00000E+00,                     1P      1
                        3.50000E+01,  5.00000E+00,                     1P      2
                        1.08527E+01,  3.44236E+01;                     1P      3
S      1G      4D      2P      3                                        T      1
```

CIRCULAR ARC				
PIXIT ref	Input Model			
	Field name	Value		
		lower	nominal	upper
1	centre point x	4.999	5.0	5.001
1	centre point y	4.999	5.0	5.001
2	start point x	34.999	35.0	35.001
2	start point y	4.999	5.0	5.001
3	end point x	10.85171	10.855271	10.85371
3	end point y	34.42256	34.42356	24.42456
4	radius	29.999	30.0	30.001
5	start angle	-0.057296	0.0	0.057296
6	end angle	78.6927	78.75	78.8073
7	swept angle	78.6927	78.75	78.8073
8	bulge factor	0.35781		
9	direction	anticlockwise		

Figure 10.6 Extract from sample executable test case.

(c) reference to specific parts of the standard(s);
(d) verdict criteria;
(e) formal language definitions of the model to exercise the test purpose.

An example of part of an abstract test case is shown in Figure 10.5. An abstract test case is processed by software to produce an executable equivalent, the corresponding example of which is shown in Figure 10.6. It includes an IGES file and a template, showing the possible alternative representations, populated with values.

10.9 Overview of the conformance assessment process

The software architecture and tools address the needs of the client, the requirements of the accreditation body, the desire for harmonization and mutual recognition. They have to be able to test all second generation processors adequately and to be embodied in a commercial service (or at least provide a basis for a commercial service). They must also contribute to a method which is also suitable for testing STEP.

The process of testing is phased and a client may not wish to participate in all phases, perhaps because he is testing the same processor on a different hardware platform or he finds difficulties with his processor that require a major rework before resubmission. The pricing of the testing is based on this phasing, so that the more phases undertaken, the more it costs.

Registration

The first action, called registration, results in the signing of the contract between laboratory and client. As well as costs and legal responsibilities, this covers the use of the results and the agreement on both sides to maintain complete confidentiality about the testing that is going to be undertaken. The testing can be undertaken either at the laboratory or on the client's premises. The choice affects only details of some of the test procedures: there is a need to agree about installation of the client's system and subsequent removal. (Accreditation demands that testing be undertaken in a secure environment accessible only by specially assigned personnel.)

The questionnaires

The first phase consists of interaction between the laboratory and the client, focused by two questionnaires, the PICS and the PIXIT. Their purpose is to elicit information about the client's processor and the environment in which it operates so that the tests applied to the particular IUT are appropriate. It has been found that the ability of the client to answer these questions, particularly the mappings between the neutral format and CAx system, depends critically on the availability of the development team. This can be a problem if the CAD system either was developed initially several years ago or has been obtained from a third party (possibly by merging of two companies), both of which circumstances are common.

In response to the answers to the questions the laboratory will select the appropriate set of abstract test cases from its library of abstract test cases and generate a set of executable test cases (ETCs).

Prevalidation

The next phase consists of the client receiving the set of executable test cases for use in-house as he wishes. The assumption is that he will wish to test his proposed IUT at his own site to ensure it will pass the formal phase of testing. If it does not process the ETCs adequately the client may wish to withdraw. Software tools for comparison and analysis are not provided as part of the conformance-testing exercise, although it is common for laboratories to provide some consultancy time as part of the conformance-testing process.

Formal phase

The final phase of conformance testing is the formal conformance testing. The laboratory is required to have a secure room and a test environment that includes computers for running software tools and well defined networking facilities for connection to the SUT. The client will be informed of the set-up but it is expected that the client's own hardware will be supplied so that the system under test is as operated under site conditions and, more importantly, as sold to the client's customers. There will thus be a need to install the system under test, to bring up the software, and be confident that it is working as the client expects, by performing a calibration test. Obviously, tests at client sites are considerably more popular; however, the calibration still needs to take place.

Carrying out of the planned tests takes place in "test time". Their exact form will depend on the specification being tested and the responses to the questions asked earlier of clients. Generally it will contain a series of tests which probe the form of the overall file (e.g. the handling of reals and strings, and the form and processing of start and terminate sections). Then the entities will be tested with each aspect having a separate test so that there will usually be a group of tests applied to each entity. Finally, for the postprocessor only, a collection of all entities in all their different forms will be processed to enable some measure of postprocessor coverage to be gained: it should be able at least to read all IGES entities, even if it cannot convert them into the CAx system.

The results must then be analyzed. For the postprocessor, the analysis compares the result of processing the file containing the entities under test with the reference results held in the abstract test case (and hence in the executable test case too). For the preprocessor, the file (which is the output from the CAD system and preprocessor) is checked for syntax conformance and is compared with the reference results. The representation supported by the client is provided with values. These are the

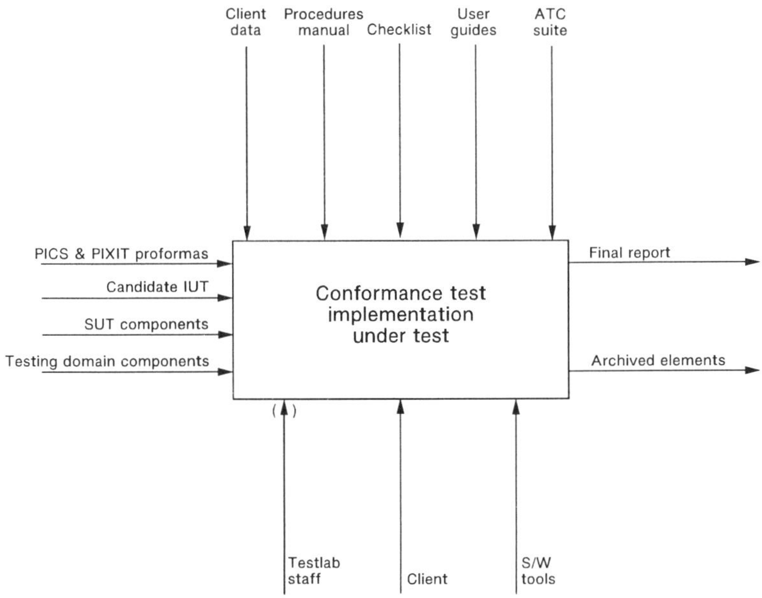

Figure 10.7 IGES conformance testing: A-0.

reference results with which the observed outcome can be compared by the person undertaking the testing.

There is a pro forma for the report, and the final report is presented to the client on completion. An independent Test Method Control Board [34] exists to which laboratory and client can appeal for advice and arbitration.

The final stage of system breakdown (i.e. removal of the system by the client) is largely a contractual issue but illustrates why a contract is important. Some retesting may be required; it may be desirable to undertake some additional tests to help to locate a particular problem identified during conformance testing. However, the space at the laboratory might be required for another processor to undergo conformance testing. In such a case, the conflicting commercial requirements are addressed in the contracts themselves.

The stages of the process shown in Figure 10.7, broken down from the activity diagram in Figure 10.8, are summarized in Figure 10.9.

10.10 CTS2

Much of the methodology has been developed in Europe within the CTS2 CADCAM (data-exchange interfaces) project [15, 35]. The partners

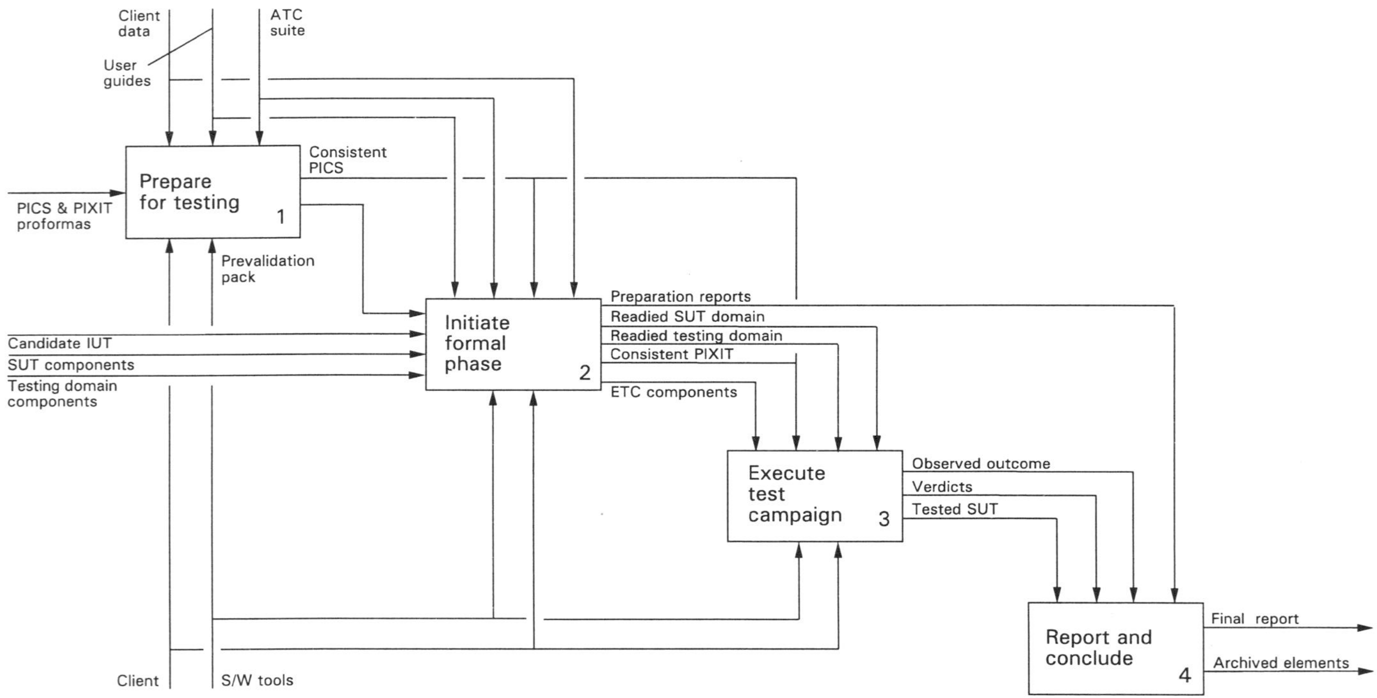

Figure 10.8 IGES conformance testing: A0.

A1: **Prepare for testing** To register the client and gather information about the processor and to deliver a prevalidation pack

A11: **Register Client** To receive the client's completed registration document which enables the test laboratory to commence work

A12: **Gather Processor Information** To collect some information about the processor to be tested for conformance and provide the client with more information about the conformance testing service

A13: **Generate & Send Out Prevalidation Pack** To generate and send a prevalidation pack to the client to enable the client to perform in-house testing prior to undertaking formal phase

A131: **Select Test Purposes** To select those test purposes which may be used for subsequent phases of testing using the PICS response

A132: **Select ATC Components** To select the ATCs prior to the generation of the corresponding ETCs which are required in the prevalidation pack

A133: **Generate Prevalidation ETC Components** To create the ETCs that are required as part of the prevalidation pack

A134: **Collate and Send Out Prevalidation Pack** To prepare and deliver the prevalidation ETCs and supporting documentation to the client to enable the client to perform in-house testing prior to undertaking formal phase

A2: **Initiate Formal Phase** To start the formal phase by receiving and checking the PIXIT and preparing for the test campaign

A21: **Receive PIXIT** To receive the completed PIXIT from the client which is an indication that the client is ready to progress to the formal phase

A22: **Check PIXIT** To analyze the remaining information provided by the client required about the processor to enable the test campaign to commence

A23: **Prepare for Test Campaign** To use the selected ATCs to generate ETCs, and to prepare the testing and the SUT domain ready for the test campaign

A3: **Execute Test Campaign** To start testing the CAx IGES processor

A31 (pre): **Perform Tests & Analyze** To run the ETCs and to analyze the IGES file generated

A311: **Create Model in SUT** To create the model in the SUT using the ETCs

A312 **Create IGES File** To use the processor of the SUT to create an IGES file for the model created in the SUT

A313: **Analyze IGES File** To check the resulting IGES file for syntax errors and to compare it with a reference IGES file

A31 (post): **Perform Tests & Analyze** To run the ETCs and to analyze the observed outcomes

A311: **Read IGES File** To read in the IGES file into the CAx system

A312: **Interrogate SUT** To examine the data-based content and/or the graphical output resulting from reading the IGES file

A313: **Analyze Observed Outcome** To compare the observed outcome with the reference results to determine whether the verdict criteria have been met

A32: **Assign Verdicts** To assign a verdict to the results obtained by running each ETC

A33: **Agree Verdicts** To agree with the client on the verdict assigned and to categorize any problems encountered during the test campaign and take action accordingly

A34: **Conclude Testing** To collect together all results and material produced during the test campaign

A4: **Report & Conclude** To document the results of conformance testing for the client (the final deliverable for the client). Also to archive all documentation, both paper and electronic, which has been generated during the conformance testing process

Figure 10.9 Breakdown of A0 (Conformance Test Implementation Under Test): IGES preprocessor or postprocessor.

were charged to ensure that the work influenced STEP and also the methods used for conformance testing worldwide, particularly in the United States and Japan.

When CTS2 CADCAM was being formulated a memorandum of agreement (M-IT-03) had been proposed and was under discussion which would have represented a great advance and was viewed with optimism by the conformance testing fraternity. This was to the effect that it would only be possible for a product to be advertised under the name by which a particular standard was known (e.g. IGES) if it had been tested for conformance. Government procurements would only be able to purchase conforming products, if there were any available. M-IT-03 is reproduced in the ECITC guide [46]. Although the memorandum was formally agreed, it was in a weakened form and has not had the desired effect. This means that the establishment of conformance testing services and their use by vendors relies on user pressure backed by understanding of the scope and possibilities of testing.

10.11 Conclusion

Definition, implementation and use of the early product data exchange standards and specifications brought several problems to light. Some could be solved in the short term, either with the next version of the specification or, in the case of IGES, with faster-track Recommended Practices which can be published more quickly but short-circuit the process. However, others required a more formal approach to definition, which could not be reverse-engineered into this second generation. In particular, the need for the standards to be testable was identified.

Much expertise was gained in testing, progressing from informal demonstrations of product data exchange working, through informal testing programmes, culminating in the availability of accredited conformance-testing services. As a result of European leadership in this area, the infrastructure for conformance testing became better understood, which would ensure reciprocal recognition of results and bilateral arrangements between laboratories for harmonized procedures.

The concepts of levels (in GKS) and subsets (in VDA-IS) provided the platform for application protocols in STEP. The "completeness" requirement – that vendors support *all* constructs in a particular subset – was also introduced into application protocols, again based on German expertise with GKS and VDA-IS.

All of these ideas were applied to and tested on the second-generation standards and specifications, but the expertise was available for the STEP development process too, which benefited enormously from the experience. This development is addressed in Part IV of the book.

Part IV

The third generation: STEP

Historically, the development of STEP (standard for the exchange of product data) started as second-generation specifications were being published in their second or even third versions. However, it is described separately in this text for two reasons. First, it is a truly international effort, based on the considerable expertise gained from the development, implementation and use of existing standards and specifications. Secondly, it was necessarily able to start afresh, without being hampered by mistakes or oversights made with an earlier version.

Part IV charts the development of STEP, documenting the key decisions taken and describing the committee structure which supports that development. Research and development projects and industrial consortia which have provided key technical effort are mentioned, including some of the centres of expertise.

STEP itself is documented as a series of parts, divided into classes. This structure is described, with the reasons why STEP has this structure, and a description (albeit brief) of some of the parts. A common theme – that of data about the British royal family – is used to link the various examples. Chapter 12 is intended to give a flavour of STEP: more details are available elsewhere [108, 94].

Part IV concludes by indicating likely future developments. These include developments of and within STEP itself, in parallel with STEP, and how the techniques with which STEP has been created can be used to make an enterprise more effective. By using STEP – either directly or indirectly – the exchange of product data in machine-readable form, and its management amongst co-operating enterprises, can become more reliable and even more commonplace than it is today.

Chapter 11

The history of STEP and associated activities

11.1 STEP initiation

The work that is now called STEP was started by the International Organization for Standardization in December 1983, primarily on the urging of the United States, France, the United Kingdom and Germany, although others had already indicated active interest. Subcommittee 4 (SC4) of Technical Committee 184 (TC184), Industrial Automation Systems, was set up to supply a Standard for the Exchange of Product Model Data. At the first meeting of SC4 (11–13 July 1984 in Washington), Resolution 1 said:

> SC4 recognises the need for a new standard for the external representation of product model data. This standard will be based upon existing data exchange initiatives including the US IGES and PDDI, the French SET, the German VDA/VDMA-FS, and the UK NEDO.
>
> Technical work will be accomplished by existing and future national projects, organizations, and resources which will be co-ordinated and monitored by the SC4 committee. SC4 will set design objectives, establish priorities, arbitrate differences, and ensure that objectives are met and consistency is maintained.

All initiatives cited above have been described in earlier chapters.

At this time, there were active representatives from the countries mentioned above and from Sweden, Japan, Switzerland and the Netherlands. At the second meeting in Frankfurt in November 1984, the name STEP was coined, suggested by the Germans, and with the warning that it had no formal status. To provide a forum for the technical work, Working Group 1 (WG1) was set up, to which the initiatives mentioned above would contribute manpower. The STEP meetings that have continued regularly and frequently since then are generally meetings of the working group(s) which are organized to coincide with those of SC4 when resolutions and other ISO formalities demand. More than twenty-five countries are currently members of SC4.

The initial meetings were spent on understanding each other's specifications and initiatives. In particular, there was a need to understand the future national plans for four activities: IGES, PDES, SET and CAD*I. PDES and CAD*I are projects which have had a large input and influence in WG1 and thus merit sections of their own, below. The developments of IGES and SET have been described already.

There were some early fears that either the United States would propose PDES, about which the Europeans knew little at that time, France would propose SET, or Germany would champion the CAD*I work as the ISO standard. The third meeting in Paris in February 1985, resolved in resolution 9:

> SC4 reconfirms its goal expressed in resolution 2: to develop as soon as possible a single international standard for exchange of product definition data to be called standard for exchange of product model data (STEP).
>
> SC4 will not concern itself with an interim solution based on any existing national standards although it recognizes that these standards will continue to be in parallel use until STEP becomes fully operational.
>
> All efforts will be concentrated to achieve a first version of STEP so that its effective industrial use can start in 1990.

At the same time resolution 10 was passed:

> SC4 agrees that, whenever existing national standards are enhanced, development should be undertaken in parallel with the corresponding STEP effort, using the same reference model to ensure that compatible concepts are used. This will cause standards to converge towards STEP and will simplify the eventual migration from the current standards to STEP.

There was agreement that:

> The US national body is designated to take the leadership role for the development of the ISO STEP standard.
>
> The US national body will direct its effort towards the development of STEP. The ongoing PDES initiation effort will be completed.

> Following completion of STEP development, the national standards and the ISO standard must be identical.

Thus the working group started to define the requirements for STEP. That it could be finished and in use by 1990 seemed optimistic to those interested in the research problems to be solved but a necessity to those pragmatists who knew industry's current problems. At the time of writing, the two sides still exist and the authors believe that STEP's common use for exchange in industry is still two years away, although it is now starting to be used for industrial exchanges.

The requirements for STEP included the objective for:

> the creation of a standard that enables the capture of information comprising a computerized product model in a neutral form without loss of completeness and integrity, throughout the life-cycle of the product.

The important points are that STEP would provide a *complete* representation of the product model, and not a graphical or visual representation. It would be developed with a methodology which incorporated

(a) reference models;
(b) formal definition languages; and
(c) a three-layer architecture (application, logical, physical).

These objectives and their technical realization are explained in Chapter 12. It is enough to say that at that stage the working group envisaged a collection of information models that captured the information that applications (e.g. draughting, AEC, electrical) and topics (e.g. surface representation, presentation) wished to communicate. These reference models would be written in an information modelling language such as NIAM [157] or IDEF1X [73] which are described briefly in Appendix C. They would comprise the application layer. Although there was software for processing these languages, the main aim of using them was to ensure that the humans defining the requirements could communicate with each other by having a common mode of expression.

The logical layer would consist of a collection of entities written in a data definition language and hence be computer-processable as well as human-readable. These entities would be sufficient to represent all those aspects in the proposed information models. The collection of entities written in EXPRESS, the data definition language designed for STEP but now in widespread use, would be the definition of the standard. This EXPRESS definition would be separated from the implementation method which in the early days was a physical file format, the physical layer of the model. The use of unambiguous context-free grammars for the definition of both the EXPRESS syntax and the syntax of the physical file format enables formal methods to be used in the development of software. However, it does not guarantee that either the definition or any supporting software is correct.

One original requirement was that the first release of STEP should have at least the functionality of IGES. The development of the geometry was split from draughting and presentation. This was a sound idea, because geometry can be used separately from draughting and presentation, as well as providing facilities for them to use. However, it resulted in difficulty in reaching agreement about the draughting information model. Not only does each industrial sector have different requirements, but the extent of the presentation model and its relationship to graphics standards (Ch. 6) also caused problems.

Since those early objectives were determined, understanding has increased, but emphasis and terminology have changed. Even so, it is fair to say that the original requirements for exchange [160] remain largely intact.

11.2 Results of the first STEP ballot

Since the mid-1980s, work has continued with frequent meetings and an increasing number of participants: people, countries and companies. There was increasing pressure to produce an initial release and frustration with the difficulty in obtaining a common level of understanding with the constantly changing population as company priorities changed. Governments committed resources so that people could do work between meetings. The national effort in each country is co-ordinated by the national standards body, contact details for which are given in Appendix B.

At the Tokyo meeting of SC4 held in December 1988, STEP was submitted for national ballot as a Draft Proposal (DP) for adoption as a Draft International Standard (so the document was known as DP 10303 STEP).

In ISO, a document is developed first in a Working Group, and has the status of a Working Draft. When there is consensus within that Working Group, it becomes a Committee Draft (or, as it was in 1988, a Draft Proposal) of the parent Sub-Committee and is distributed to nations participating in the SC for a three-month or four-month ballot. If successful, it becomes a Draft International Standard of the parent Technical Committee, and is distributed to all nations participating in ISO for a six-month ballot. If successful in this round, it becomes a full International Standard. Of course, a document may be rejected at any stage, in which case it is returned to the committee or group which put it forward for ballot so that the issues raised against it can be considered and the document amended subsequently.

Table 11.1 Contents of the Tokyo Draft Proposal.

Title	Clause
scope	1
normative references	2
definitions	3
introduction	4.1
resource schema	4.2
types and functions	4.3
miscellaneous resources	4.4
geometry	4.5
topology	4.6
shape representation	4.7
features	4.8
shape representation interface	4.9
tolerancing	4.10
materials	4.11
presentation	4.12
product life cycle	4.13
applications	4.14
product manifestion	4.15
product structure	4.16
AEC applications	4.17
ship models	4.18
electrical applications	4.19
analysis applications	4.20
data transfer applications	4.21
conformance	5
EXPRESS	A
physical file structure	B
mapping to physical	C

The structure of the Tokyo DP is shown in Table 11.1. The integrated product information model (or IPIM), which comprises Clause 4 of the Tokyo DP, consists of three major sections (resources, life-cycle and applications), each of which contains models written in EXPRESS. Of these three sections, the life-cycle was not developed at all and there was disquiet about some of the resources (particularly features and tolerances) and about the applications (particularly the product structure configuration model and the electrical models). Also included was a set of miscellaneous resources whose scope was the whole integrated product information model (e.g. items such as time, person, name and other general-purpose types and functions). Although the IPIM formed the majority of the Tokyo DP, the other clauses were also important, particularly the normative annexes describing EXPRESS, the physical file structure, and the mapping from EXPRESS to the physical file.[4]

Because pressure for wider review was becoming great, the work was submitted as a single document with EXPRESS and the physical file format as part of the standard [106, 172, 53, 141, 130, 2, 3, 4]. Personal preferences and experience had prevented the community from being able to agree on a single information-modelling language, but the information models written in NIAM and IDEF1X were added as informative annexes, as “road maps” to help the reader [39].

The intention of the working group in submitting this document was to expose the work widely and to obtain constructive comment. The results of the national ballots of STEP were presented at the Frankfurt meeting in June 1989. Although STEP was presented as a single document, informal votes were cast for each clause. The overall vote was to disapprove, but a large amount of exposure was achieved. For example, in the UK, the AMT/4 Committee of BSI organized meetings to consider the various clauses and annexes. These brought together experts from industry and academia and, as well as providing comments on the proposals, they increased mutual understanding of the current engineering data management problems. A large number of comments were communicated to the working group and many improvements were made as a result.

The major changes were the division into parts and the embracing of the concept of application protocols on which the IGES Testing Project was already at that time working. These arose naturally because of the different levels of maturity of the different models. Further, it would help

Table 11.2 The STEP committees.

Committee	Convener
TC 184 Industrial Automation Systems and Integration	François de Belenet (France)
SC 4 Industrial Data and Global Manufacturing Programming Languages	Brad Smith (USA)
WG 1 (disbanded in June 1990)	Jerry Weiss (USA)
WG 2 CAD standard parts	Gerd Ehinger (Germany)
WG 3 Product modelling	Barbara Warthen (USA)
WG 4 Qualification and integration	Yuhwei Yang (USA)
WG 5 STEP development methods	Bill Danner (USA)
WG 6 Conformance testing procedures	Sheila Lewis (UK)
WG 7 Implementation specifications	Jan van Maanen (UK)
WG 8 Industrial manufacturing management data	Albert Colin (France)
JWG 9 Electrical and electronic product model data	vacant (USA)
Editing Committee	Nigel Shaw (UK)
Strategic Planning Advisory Group (SPAG)	Jean-Pierre Letouzey (France)
Project Management Advisory Group (PMAG)	Neal Laurence (USA)
Ad hoc Application Protocols (APs) Group	Mark Palmer (USA)

to prevent vendor implementation of uncontrolled subsets, although there was never any intention that the whole of STEP would be implemented by a single vendor. Consequently, out of the Frankfurt meeting came the structure shown in Figure 12.1 which shows the classes into which STEP is now divided. The parts under development are shown in Table 12.1 with their status at the time of writing.

After Frankfurt, it was agreed that the initial release should contain at least one application protocol and the functionality of IGES. The size of both the working group and the standard were recognized as becoming unmanageable and there were fears about the lack of integration between the resource models and about the technical quality as well as uniformity of presentation. A reorganization into Working Groups rather than the *ad hoc* groups which had grown up was undertaken after the Gothenburg meeting in June 1990. The structure is shown in Table 11.2, with the conveners at the time of writing, although a new structure is about to be put into place.

11.3 PDES initiation

PDES was originally the acronym "Product Data Exchange Specification" for a research project started by the IGES Organization in 1984. Its intention was to develop and use a rigorous methodology for the creation of product data exchange specifications. It followed the PDDI project and wished to understand the information that supported design and manufacture. This would facilitate the integration of the two phases by enabling the understanding of the levels of abstraction more fully than in the past, when the islands of automation had been interfaced rather than integrated. Wilson [166] explains the PDES objectives in terms of the relationship between data and information and levels of detail and higher abstractions that come with better understanding.

The IGES Organization became the IGES/PDES Organization (or IPO) and the two projects continued in parallel. After the report of the initiation phase in May 1986 [143], its original status – that of research – was interpreted by some from the United States as that PDES would be the basis for the international standard. For several years, the ISO work progressed and although those involved knew that STEP and PDES were the same (as ratified in Resolution 9, shown in Section 11.1), it was a constant battle to prevent divergence and to ensure that the wider community realized this too. Eventually, in 1990, IPO announced that the acronym PDES was in future to stand for "Product Data Exchange using STEP"; that is, it became an initiative rather than a specification.

IPO now has more than 600 individuals representing over 200 organizations from industry, government and academia giving their time voluntarily, which is the way in which standards organizations work. These individuals belong to many technical committees and interest groups working on three concurrent projects, IGES, PDES and Testing, which report as shown in Figure 3.4. As in the beginning of IGES, IPO is still supported strongly by NIST, which provides an office and support and undertakes tasks for both ANSI and ISO.

The National Computer Graphics Association (NCGA) is the administrator for the IPO. It co-ordinates all meetings and events for the organization, and assists with publicity and promotion. It is responsible for education and information including IPO document distribution, seminars and a national IGES user group.[5]

11.4 CAD*I

CAD*I is the name by which ESPRIT project number 322 is known [86]. This research project started in late 1984, about the same time as the IPO PDES Initiation project. The partners in the project were as follows:

Kernforschungszentrum Karlsuhe (Germany)
BMW (Germany)
Cranfield Institute of Technology (UK)
Katholieke Universiteit Leuven (Belgium)
Leuven Measurement and Systems International (Belgium)
Rutherford Appleton Laboratory (UK)
Universität Karlsruhe (Germany)
GfS (Germany)
Disel (Spain)
Cisigraph (France)
NEH Technology (Denmark)
Danmarks Tekniske Højskole (Denmark)

It lasted five years and its objectives were to examine and demonstrate new technologies for product data exchange, particularly of surfaces and solid geometry, finite-element data and dynamics for robotics. It finished in October 1989, having made a significant contribution to STEP in the areas listed above and also in the application of a data definition language, influenced by PDDI's specification language proposal, DSL. CAD*I produced its own high-level data language (HDSL) which it used throughout and thus gained experience that helped to form EXPRESS. It also had a significant influence on the physical file structure and introduced a structuring mechanism (called "scope") within the file.

The very valuable results of CAD*I are available in book form [132, 148, 133, 135, 121, 60, 61], which detail the representations and describe the experiments in data transfer between solid modelling systems which were demonstrated at the final review.

11.5 PDES Inc.

PDES Inc. is a joint industrial and government effort formed in 1988 to accelerate the development and implementation of PDES (STEP) [168]. It is supported by member companies with cash and manpower. Those companies contributing to PDES Inc. at the end of its first year are shown as follows:

Boeing	General Dynamics
General Electric	Grumman
Lockheed	McDonnell Douglas
Northrop	DEC
FMC	IBM
LTV	Prime/Computervision
Rockwell	Martin Marietta
General Motors	NIST (as Government Associate)

Although NIST provides large and excellent support for IPO, the amount of direct government money (as opposed to that from procurement contracts) was not proportionately as large as in Europe. Consequently, the largely voluntary effort was dependent on company priorities, which PDES Inc. intended to balance. The initial fear was that PDES Inc. saw STEP (or PDES, as some persisted in calling it) as an American effort and would not respect the international open standards-making process. Initial fears were fulfilled with PDES Inc. holding closed meetings at IPO and ISO meetings and showing a lack of understanding of the standards-making process. As the manpower committed by the companies to PDES Inc. became experienced and those already involved in STEP were assigned to PDES Inc. by their companies, this situation has improved, so that the fear is no longer justified. The change of meaning of PDES, from a specification to an initiative, helped considerably. There is still a tendency to push proposals for application protocols before the integrated resources are in place to support them. This is because of the perceived company need to have (for example) product structure, tolerances and features to support integration of design and manufacture as well as data exchange.

11.6 CALS

Continuous Acquisition and Life-cycle Support (CALS) is a DoD and industry strategy to integrate digital technical information for system acquisition, design, manufacture and support [66, 44]. The strategy names several forms of data exchange as crucial to this strategy, and requires the use of certain standards to effect exchange in each particular area:

- structured documents: SGML;
- raster (scanned) images: CCITT Group IV;
- two-dimensional vector graphics: CGM;
- product data: IGES, EDIF and VHDL in the short term and STEP in the longer term.

So-called "CALS compliance" will require the use of these standard formats for government contract deliverables. As a strategy it is necessary and laudable and it has received a lot of publicity. In substance there is currently very little because processors are not adequate and the conformance testing that ensures they can be used effectively is not in place for all of the standards. However, the CALS initiative is probably doing more good than harm by providing education and fostering action. The problem only comes if people believe that because the DoD mentions these standards it means there is no more work to be done.

In the DoD, CALS Phase II has shifted its emphasis from the delivery of digital information from industry to government and to the sharing of information between industry and government. This can only be done through an integrated database and data models that can be accessed by a variety of heterogeneous computer systems. STEP is seen as providing the methodology by which product data can be shared between these systems. CALS Phase II strategy relies on the completion of STEP and the implementation and use of STEP by all parties.

The CALS Test Network is a confederation of participants from industry and government. It covers both conformance testing and user application testing. Lammers [87] lists the tests completed and the software tools available, and gives an overview of the tests themselves.

11.7 CADEX

CADEX is an ESPRIT Phase II project to support STEP by writing processors for geometry and finite-element systems. It was designed originally to take the successful CAD*I prototypes and make them into products to help European CAD vendors. The partners are the following:

GfS (Germany)	Caddetc, University of Leeds (UK)
BMW (Germany)	FEGS (UK)
Hewlett-Packard (Germany)	Fiat (Italy)
PROCAD (Germany)	Siemens (Germany)
SI (Norway)	DnV (Norway)
Norsk Data (Norway)	ITALCAD (Italy)
ERDISA (Spain)	ISYKON (Germany)

The predominance of Europe-based vendors is clear.

The original intention was to use the application protocols being developed in STEP – particularly the geometry-based ones (parts 204,

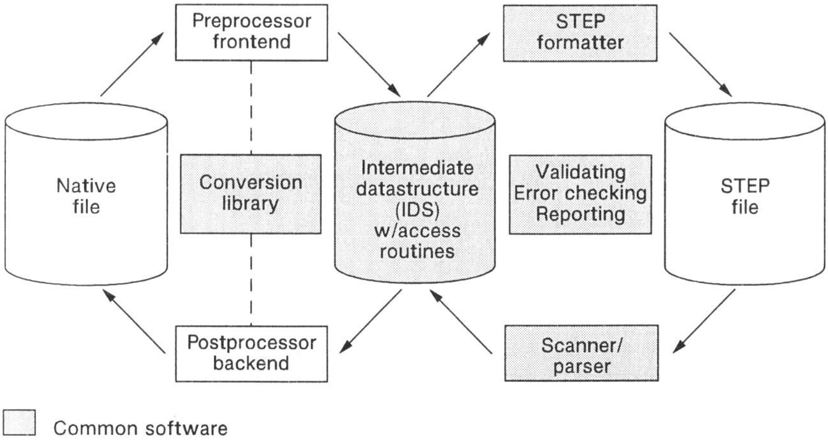

Figure 11.1 CADEX software architecture.

205 and 206) – and use these as the basis for processor development. However, these application protocols were not sufficiently stable for this, so the CADEX project provided crucial manpower to bring them to a state where they could be used. This task was in addition to the software development, which defined a software architecture, shown in Figure 11.1, and produced a software toolkit for use by the vendors in developing their own STEP processors. The architecture included model access software, similar to that produced by the PDDI project, and destined to become a STEP implementation method known as the SDAI. Testing was also undertaken, and demonstrations were given of product data being exchanged successfully between many of the systems.

11.8 European projects

As well as the European-funded projects mentioned already (CAD*I, CTS2 and Cadex), NIRO, IMPPACT and Neutrabas (completed in 1992) and VIMP (completed in 1993) made significant inputs to STEP development.

The European standards committee (CEN/CENELEC) contributed to certain parts of STEP to make them suitable for European standardization. This included effort on Part 201 and on standard parts.

With the ESPRIT III round of research and development projects, there is a considerable amount of technical expertise in Europe which is contributing to STEP. The projects are too numerous for us to describe each of them; but they include the following:

- Maritime: integration of ship-building activities;
- Pisa: framework for information modelling;
- Prodex: data access interface software for mechanical design;

- Consens: concurrent engineering;
- Fires: integration of design with features, analysis and process planning for complex three-dimensional components;
- 3Dfashion: cooperative CAD tools in three dimensions for the apparel industry;
- Interrob: on-line robot control using neutral programming interfaces and simulation;
- Processbase: data to support design, construction, operation and maintenance of process plants;
- Atlas: integration of large-scale engineering projects (building and civil);
- Combine: integrated systems for building design;
- PI-STEP: process industries;
- CIMSteel: constructional steelwork.

Further details are given by Gielingh [55], and a database of information is maintained by the European Union's PDTAG (see Section 11.11) at the University of Madrid.

11.9 Electrical and electronic applications

This application area deserves a section on its own because one of the disappointments with STEP has been the failure to encompass and learn from the good work, described in Chapter 6, that the electronics industry has undertaken. Although the industry has chosen to avoid the formal standardization process and rivalries between the large semiconductor companies are apparent, the work has continued. There have been difficulties in persuading those in the EDIF organizations and the IGES/PDES Organization to come together at the technical level. A step forward in willingness was made at the SC4 Paris meeting in January 1990.

In addition the BSI promoted a project [38] to compare STEP and EDIF from the point of view of electronic coverage. The overlap was found to be small, namely, use of the STEP geometry and presentation to simulate the graphic elements by which schematics are presented in EDIF. However, that project attempted for the first time to represent some of the EDIF functionality in EXPRESS. The result is available from *uk.ac.man.cs.edif* using anonymous ftp. The shortcomings of EXPRESS in representing the relationships, such as logical connectivity, between design objects that are important in electronics, electrical engineering and hydraulics have been demonstrated and led to the enhancement of EXPRESS version 2.0. The latest version of EDIF [48] also includes an EXPRESS model of the specification.

Because of the unsatisfactory level of co-operation in this and other areas, an organization was chartered, in May 1990, to facilitate more effective use of digital representative standards (DRS) by providing a forum for the co-ordination, planning and guidance of standards developers and approvers. Formed under the sponsorship of the ANSI Industrial Automation Planning Panel (IAPP), the organization's initial focus is on electrical/electronics product data. Realizing the cost and confusion caused by contradictory, overlapping or redundant efforts in standardization, the effort is an attempt to integrate a network of standards which can smoothly support the definition of products across their life-cycles in digital form.

11.10 Centres of expertise

In the 1980s, centres of expertise in product data exchange and management began to develop at sites of existing initiatives or research project partners. In the UK, the NEDO work and research in the Department of Mechanical Engineering at the University of Leeds were instrumental in the establishment of the CADCAM Data Exchange Technical Centre (CADDETC) in 1986. This was funded by research and development projects and several industrial companies, so that it would address both immediate and longer-term problems with product data exchange. In conjunction with the University of Leeds, at one time it provided over half of the editors for the parts in the initial release of STEP. It now provides consultancy, seminars and software. As well as those given at Leeds, the Universities of Manchester and Newcastle, and South Bank University provide seminars and courses on various aspects of product data.

In France, similar centres were established by Association GOSET, whose participants had played a leading role in the development of SET, and at ARMINES/CPAO (formerly IRPL), one of the partners in the CTS2 project described in Chapter 10. In Germany, ProSTEP was established in the eastern part of Berlin; the first phase of work has been completed, and at the ISO meeting held in Berlin in October 1993, a non-profit association and a commercial corporation were launched to undertake the second phase. In the Netherlands, the involvement of TNO in STEP and several ESPRIT-funded research projects is one focus; another is the newly established CALS centre. In the Iberian peninsular, interest is focused at the Universities of Madrid and Lisbon, and at UNINOVA (also in Lisbon). In Scandinavia, SI in Oslo is very active, and work continues at another of the CTS2 partners, IVF in Göteborg.

Many other universities and laboratories in Europe also have expertise in these areas. However, apart from ProSTEP in Berlin, which was established as such, there are no truly national centres. There are also other strong centres of excellence in Germany.

The Japanese STEP centre was established by Nicograph in Tokyo, and continues to provide support for STEP development, education and implementation. The 1993 Nicograph exhibition included a demonstration of STEP processors exchanging draughting information defined by Part 201.

In the United States, both NIST and PDES Inc. provide a focus in this area. NIST (formerly NBS) has a track record dating back before the first version of IGES, and PDES Inc. has been one of the major contributors to STEP, as described earlier in this chapter.

More recently, a centre has been established in China. There is also growing interest in Singapore, Australia and Korea. Representatives of these centres of expertise necessarily meet at the regular ISO meetings, and keep one another up-to-date with their activities.

Within Europe, there are regular meetings of EPDEN – the European Product Data Exchange Network. Representatives from TNO, GOSET, ARMINES/CPAO, SI, IVF, the University of Leeds and CADDETC (jointly), ProSTEP and UNINOVA meet to discuss joint initiatives, possible future collaborative projects, and to share news from the many projects in which they are involved individually. It is important to reduce duplication of effort which could happen only too easily, given the number of projects and initiatives.

11.11 Industrial and international co-ordination

The Product Data Technology Advisory Group (PDTAG) was established in early 1992 under the auspices of ESPRIT CIME. Its objective is to give advice and recommendations on European developments in product data technology to industry, the research community and the European Union.

It holds regular concertation meetings, in which representatives of the various European-funded research projects present their results, an annual conference (Product Data Technology Days), gives seminars and courses on all aspects of product data, and produces a regular newsletter. Its membership is drawn from both academia and industry, as follows.

Reiner Anderl (TH Darmstadt)
Alain Bézos (GOSET, France)
Albert Colin (Adepa, France)
Max Engeli (ETH Zürich, Switzerland)
Fikry Garas (Taylor Woodrow Construction Holdings, UK)
Wim Gielingh (CAP VOLMAC, Netherlands)
Carlos Guedes-Soares (Universidade Técnica de Lisboa, Portugal)
Jochen Haenisch (SI-SINTEF, Norway)

Lothar Heggmair (BMW, Germany)
Stuart Lord (EPISTLE, UK)
Paul Maebe (EC)
Jon Owen (The University of Leeds, UK)
Horst Nowacki (TU Berlin, Germany), Chairman
Guus Ranke (Phillips, Netherlands)
Alain Roumiguier (Matra, France)
Francisco Sastrón (UP Madrid, Spain)
Keith Shaw (Pafec, UK)
Nigel Shaw (PROSTEP, Germany)
Denis Vandorpe (University of Lyon, France)
Johan Vesterager (TU Denmark)
Günther Zintl (Siemens, Germany)

In the United States, the US Product Data Association (USPRO) provides management and strategic direction for organizations engaged in research, development, implementation and testing of standards and specifications for the exchange and sharing of product information. It is administered by the NCGA, and has about sixty members drawn from industry, government and academic establishments. The National Initiative for Product Data Exchange (NIPDE) is an industry-led, government-facilitated organization established to accelerate product data exchange development and implementation. This organization is administered by the US Department of Commerce at NIST.

Further industry-specific initiatives include the Automotive Industry Action Group (AIAG) and Process Data eXchange Institute (PDXI) in the United States, and the Process Industries STEP project PI-STEP in the UK.

11.12 Conclusion

STEP benefited from the experience gained from the development, implementation and use of the second-generation product data exchange standards and specifications. Collaborative research and development projects, national initiatives, the network of national centres, and action groups and projects related to particular industrial sectors have all contributed to the spectrum of knowledge which has created STEP. One of the tangible results of this expertise – the STEP standard, including its structure and contents – is described in the next chapter.[6]

Chapter 12

A brief description of STEP

12.1 The structure of STEP

The second generation of product data exchange standards and specifications tended to confuse the requirement of a particular application with both its fulfilment as a particular entity or group of entities and the realization on a physical medium. For example, IGES provides many types of finite elements, which are described in terms which a specialist in that discipline would understand. However, in the same section, it also describes in which particular column in the physical file certain information should appear, and what values may be defaulted. Whereas this information is required by implementors of the standard, it is of little relevance to a reader who wishes to determine if it fulfils particular requirements in finite elements.

This blurring causes a number of problems in itself. First, the requirements are often not stated explicitly. Consequently, it is difficult to perceive if the entities provided actually fulfil the needs of a particular application. Secondly, the nature of the physical file may be taken into consideration when defining the logical content of an entity, when what should be the main issue is fulfilling the application requirement. If the form of the physical file is determined separately, this does not become an issue. Thirdly, structures which are common across applications but which may be identified by different terms should be shared, in order not only to make the standard more efficient but also to allow sharing of product data across different disciplines. (For example, a three-dimensional Cartesian point, a finite-element node and a vertex in a boundary-representation solid model are at least similar, and may share the *x*-, *y*- and *z*-co-ordinates denoting a position in three dimensions.) Fourthly, if the use of an entity is designated as being only in a particular context, this helps vendors to determine what set of entities they need to implement, rather than providing an unstructured set from which they will pick and choose on an *ad hoc* basis, resulting in problems which have been described earlier.

At the time that STEP was initiated, a three-layer architecture was adopted, based on the ANSI/SPARC model [150]. The layers in STEP were

(a) application;
(b) logical;
(c) physical.

This immediately provided a structure in which the various application experts could document their requirements, the logical layer experts could provide an integrated set of entities to fulfil all of these requirements, and implementation experts could concentrate on providing an efficient format irrespective of what information it would contain. This was reflected in the subgroups of TC184/SC4/WG1 set up in 1984.

When STEP was balloted as a Draft Proposal at the end of 1988, a single document was circulated. However, each country was asked to provide an informal vote on each of the individual clauses in addition to the official vote on the entire document. The results of the vote have been described by Wilson [169]. It was not surprising that the vote was to reject the document: all work had been included, and clauses were at different stages of development and maturity. However, there was consensus on the core of STEP, such as EXPRESS and the physical file, and on the geometric facilities it provided. What was missing was a subsetting mechanism and a framework. As a result, the document was divided into a number of classes and parts. Parts would belong to a particular class and could progress at different rates, provided that at any one time a sensible set of documents was advanced to draft international standard. Clearly, it would not be useful to have a series of models defined in EXPRESS if neither EXPRESS nor the framework for them were available.

This also enabled the adoption of a concept which was being developed in the IGES/PDES organization, specifically by the Application Validation Methodology Committee, under the leadership of Mark Palmer. This concept was known as the application protocol, and was based on the application profiles which were being developed for OSI. An application protocol would take entities which existed in the logical layer, now called the set of integrated resources, and *interpret* them for use in a particular application. This interpretation involved specialising the entities or adding constraints to the original general-purpose entities to make them fit for a specific purpose in the application protocol. (Although there are several types of B-spline curve and surface, a particular application may allow only the rational form, for example.) The application protocol would provide the framework for a STEP implementation, because it was self-contained.

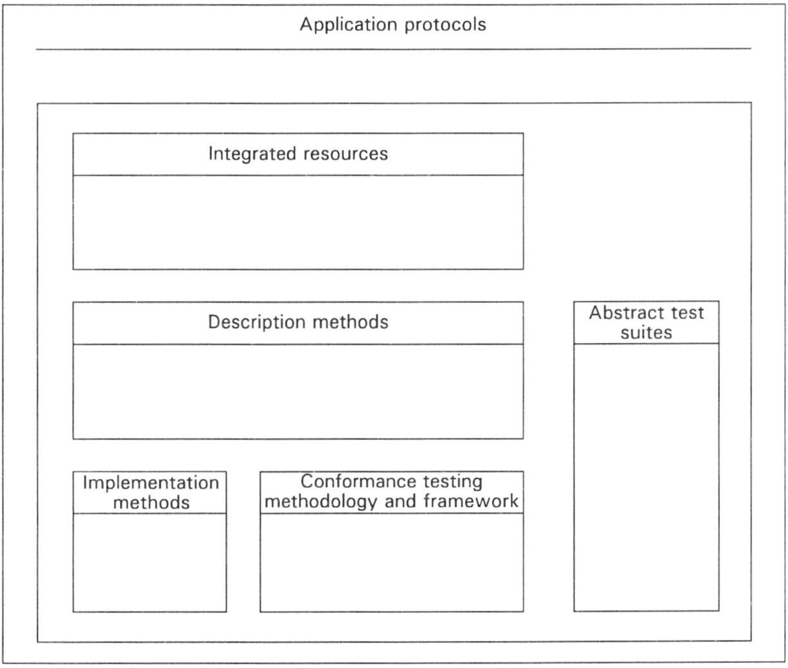

Figure 12.1 The STEP classes.

The resulting structure of STEP is shown in Figure 12.1. The three-layer architecture is still present, as application protocols, integrated resources and implementation methods. Description methods initially provided the EXPRESS language, used to define the integrated resources and application protocols. Work on conformance testing had been undertaken as part of STEP since the start of the project, and was acknowledged by the allocation of an entire class of parts; abstract test suites were added as a class in 1991.

The principal relationships between the various classes are as follows:

(a) Formal *description methods* are used in the definition of *integrated resources*.
(b) *Application protocols* are developed for a particular application context using the *integrated resources* and *description methods*.
(c) An *application protocol* is combined with an *implementation method* to form the basis of a STEP implementation.
(d) A STEP implementation is tested for conformance to the standard using the *conformance testing methodology and framework* and the *abstract test suite* associated with the *application protocol*.

Each of these classes is now described in turn, with examples. The theme used to link the examples is data concerning the British royal family; part of the family tree is shown in Figure 12.2.

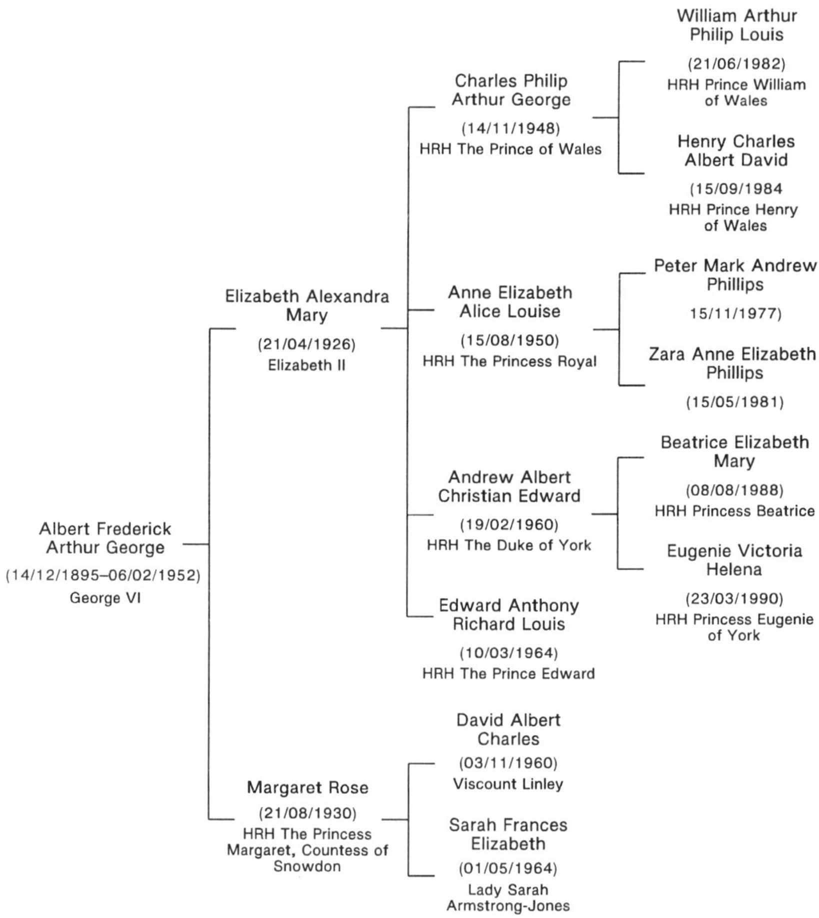

Figure 12.2 Part of the British royal family tree.

The parts which comprise STEP are shown in Table 12.1 with their status at the time of writing.

Notes on summary of parts: The critical parts (i.e. those intended for the initial release of STEP), as determined at the Paris WG and Gothenburg SC4 meetings in 1990, are marked with a †; those indicated form the original set of nine documents. Those marked with a ‡ are now also included in the initial release. The class of abstract test suites is not shown; parts are numbered one hundred higher than their corresponding application protocol, provisionally.

Each of the sections in this chapter is intended to give only a flavour of each topic: a reader who requires further details should consult the standard documents themselves. More helpfully, in the case of EXPRESS, a textbook is available already [131].

Table 12.1 Status of STEP parts.

part	title	status	comments
1	Overview and fundamental principles †	IS	
	Description methods:		
11	The EXPRESS language reference manual †	IS	
12	The EXPRESS-I language reference manual	CD	SC4 N284 ballot closes 27/06/95
	Implementation methods:		
21	Clear text encoding of the exchange structure †	IS	
22	Standard data access interface	CD	SC4 N280 ballot closed 19/03/95
	Conformance testing methodology and framework:		
31	General concepts †	IS	
32	Requirements on testing laboratories and clients	CD	SC4 N227 ballot closed 03/03/94
33	Abstract test suites	CD	SC4 N307 ballot closes 30/08/95
34	Abstract test methods	WD	SC4 /WG6 N81+ dated 17/03/95
	Integrated resources:		
41	Fundamentals of product description and support †	IS	
42	Geometric and topological representation †	IS	
43	Representation structures ‡	IS	
44	Product structure configuration ‡	IS	
45	Materials	CD	SC4 N221 ballot closed 21/01/94
46	Visual presentation †	IS	
47	Shape variation tolerances	CD	SC4 N222 dated 21/09/93
48	Form features	WD	SC4 /WG3 N102 dated 02/01/92
49	Process structure, property and representation	CD	SC4 N223 dated 21/09/93
101	Draughting †	IS	
102	(Ship structures)	WD	now deleted
103	Electrical applications	WD	
104	Finite element analysis	CD	SC4 /WG3 N263x dated 09/09/94
105	Kinematics	DIS	ballot closes 22/12/95
	Application protocols:		
201	Explicit draughting †	IS	
202	Associative draughting	CD	SC4 N226 ballot closed 01/03/94
203	Configuration controlled design ‡	IS	
204	Mechanical design using boundary representation	CD	SC4 N271 ballot closed 11/03/95
205	Mechanical design using surface representation	CD	SC4 N272 ballot closed 11/03/95
206	Mechanical design using wireframe representation	WD	SC4 /WG3 N114 dated 12/06/92
207	Sheet metal die planning and design	CD	SC4 N220 and N234 dated 21/09/93)
208	Life cycle product change process	CDC	
209	Design through analysis of composite and metallic structures	CDC	SC4 /WG3 N233 dated 09/09/94
210	Electronic printed circuit assembly, design and manufacture	CD	SC4 N258 ballot closed 07/10/94
211	Electronics test, diagnostics and remanufacture	WD	
212	Electrotechnical plants	CDC	SC4 /JWG9 Nxx-94 dated 04/07/94
213	NC process plans for machined parts	CD	SC4 N259 ballot closed 07/10/94
214	Core data for automotive design processes	CDC	SC4 N229 period closed 03/02/94
215	Ship arrangement	WD	
216	Ship moulded forms	WD	SC4 /WG3 N290
217	Ship piping	WD	
218	Ship structures	WD	SC4 /WG3 N302 dated 10/05/94
219	Dimensional inspection process planning for coordinate measuring machines using tactile and video sensors	WD	
220	Printed circuit assembly manufacturing planning	CDC	SC4 N262 period closed 26/11/94
221	Functional data and schematic representation for process plants	CDC	SC4 N362 period closed 28/02/95
222	Exchange of product definition data from design engineering to manufacturing engineering for composite structures	WD	
223	Exchange of design and manufacturing product information for cast parts	WD	
224	Mechanical products definition for process planning using form features	CDC	SC4 N273 period closed 11/02/95
225	Structural building elements using explicit shape representation	WD	
226	Ship mechanical systems	WD	SC4 /WG3 N290
227	Plant spatial configuration	CDC	SC4 N368 period closed 28/02/95

12.2 Description methods

EXPRESS is a textual language used for the definition of conceptual schema [82]. It allows entities to be defined with their attributes, and relationships between them, and includes provision for generalization and specialization. Although EXPRESS was developed initially for STEP, it is now being used outside that project (e.g. the EDIF work cited in Chs 6 and 11).

EXPRESS enables the structure of information to be specified. Such a structure can be depicted graphically using EXPRESS-G, and an instance (i.e. a model with values) specified using EXPRESS-I.

Two of the main STEP requirements for EXPRESS were that it be human-readable and computer-processable, both of which allow communication and reduce the problems of ambiguity discussed earlier. However, it should be stressed that an EXPRESS schema by itself is not sufficient to define the standard: explanation in English is required to support the EXPRESS definitions and to explain the meaning of the names chosen for the entities and attributes.

An example of an EXPRESS schema is provided in Figure 12.3. It is delineated by the keywords `SCHEMA` and `END_SCHEMA` and contains two type declarations, three entity declarations, a rule and a function (the last two of which are incomplete). EXPRESS includes provision for defining constants and procedures, but these are not shown in this example.

EXPRESS provides a number of simple types as part of the language, such as integer and string, and constructs for aggregation, such as array and set. The first type declaration, named `date`, defines a structure which is an array of three integers. This type is now available for use elsewhere in the schema. The second type declaration is an enumerated type called `hair_type`, which gives the predefined set of values to which an instance of that type may be assigned.

Each entity contains some explicit attributes, which have a name and a type. The entity `person` has explicit attributes named `first_name`, `last_name`, `title`, and so on; the type of each attribute appears after the colon. These types are either the EXPRESS simple types, aggregations, or types defined in this schema. Thus, `last_name` is a string and `birth_date` and `hair` use the types defined by the two type declarations at the start of the schema. The `first_name` of a person is a set of strings with at least one element; for each `person`, each string must be unique (because the type is a set) and there is an unlimited number of first names which can be given to a `person`. Similarly, a `person` may have `children`, but in this case the set may be empty. Note that each of the `children` is also a `person`.

Some explicit attributes may be designated as optional – for example, a `person` may not have a `title` or a `death_date`. This means that instances of a `person` can include values for these attributes, or not.

```
SCHEMA royal_example;

TYPE date = ARRAY [1:3] OF INTEGER;
END_TYPE;

TYPE hair_type = ENUMERATION OF
     (fair, brown, black,
      red, grey, bald);
END_TYPE;

ENTITY person SUPERTYPE OF (ONEOF(female, male));
  first_name : SET [1:?] OF STRING;
  last_name  : STRING;
  title      : OPTIONAL STRING;
  birth_date : date;
  death_date : OPTIONAL date;
  children   : SET [0:?] OF person;
  hair       : hair_type;
DERIVE
  age : INTEGER := years(birth_date);
  alive : BOOLEAN := NOT EXISTS(death_date);
INVERSE
  parents : SET [0:2] OF person FOR children;
END_ENTITY;

ENTITY female SUBTYPE OF (person);
  husband     : OPTIONAL male;
  maiden_name : OPTIONAL STRING;
WHERE
  w1: (EXISTS(maiden_name) AND EXISTS(husband)) XOR
      NOT EXISTS(maiden_name);
END_ENTITY;

ENTITY male SUBTYPE OF (person);
  wife : OPTIONAL female;
END_ENTITY;

RULE married FOR (female, male);  (* to be written *)
(* checks pairwise relationship between spouses *)

FUNCTION years(past : date): INTEGER; (* to be written *)

END_SCHEMA;
```

Figure 12.3 Example of an EXPRESS schema.

An entity may also have derived attributes. This information would not be stored explicitly in an instance, but could be derived when required: it is auxiliary or secondary in nature. Thus, the `age` of a `person` can be derived from the `birth_date` using the function `years` (assuming that today's date is known and any `death_date` is taken into consideration). Similarly, whether a `person` is `alive` can be deduced from the presence of a value for the `death_date`.

An entity may have inverse attributes, which provide a facility for modelling existence-dependence. If `parents` had been modelled as an explicit attribute, an instance of a `person` would have had to include information concerning their parents, each of which would also be a `person`. In turn, the parents would include details of their parents, and so on. By using the inverse attribute, the relationship is reversed: in this example it states that each person must be used in the role of no more than two other persons' children (i.e. cannot have more than two parents). The example does not require all biological parents to have been instanced: the constraint in that case would have been `2:2` (exactly two): it allows each `person` of interest to be instanced, but others not to be included.

The mechanisms for generalization and specialization, the supertype and subtype, are now considered. All information associated with `person` applies to *all* instances of a `person`; however, a `person` is either a `male` or a `female`, each of which has particular information associated only with that particular sort of `person`. A `male` has an explicit attribute named `wife`, but this is optional. Similarly, a `female` may have a `husband`, as an optional `male`, as well as an optional string for her `maiden_name`. If an instance of a `male` or `female` is created, then all of the explicit attributes from its supertype (a `person`) are inherited: a `female` has a `first_name`, `last_name`, `title` and so on, as well as a possible `husband` and `maiden_name`.

The final aspect of EXPRESS shown in this example is the constraint mechanisms. The entity `female` has a local rule, given by the condition after the `WHERE` keyword. This states either that values for both her `maiden_name` and `husband` are provided, or that neither are: she cannot have a value for one without the other. A global rule `married` ensures that each person is married to only one other.

Other facilities are available in EXPRESS, but space does not allow them to be presented here. The example includes the major constructs and is sufficient to provide a flavour of the language and to drive further examples later in this text.

Although EXPRESS is a textual language, it does have a graphical form, EXPRESS-G, in which many of its constructs can be rendered pictorially. This enables readers to be able to assimilate the structure of a schema more quickly, noting the entities and types which are of

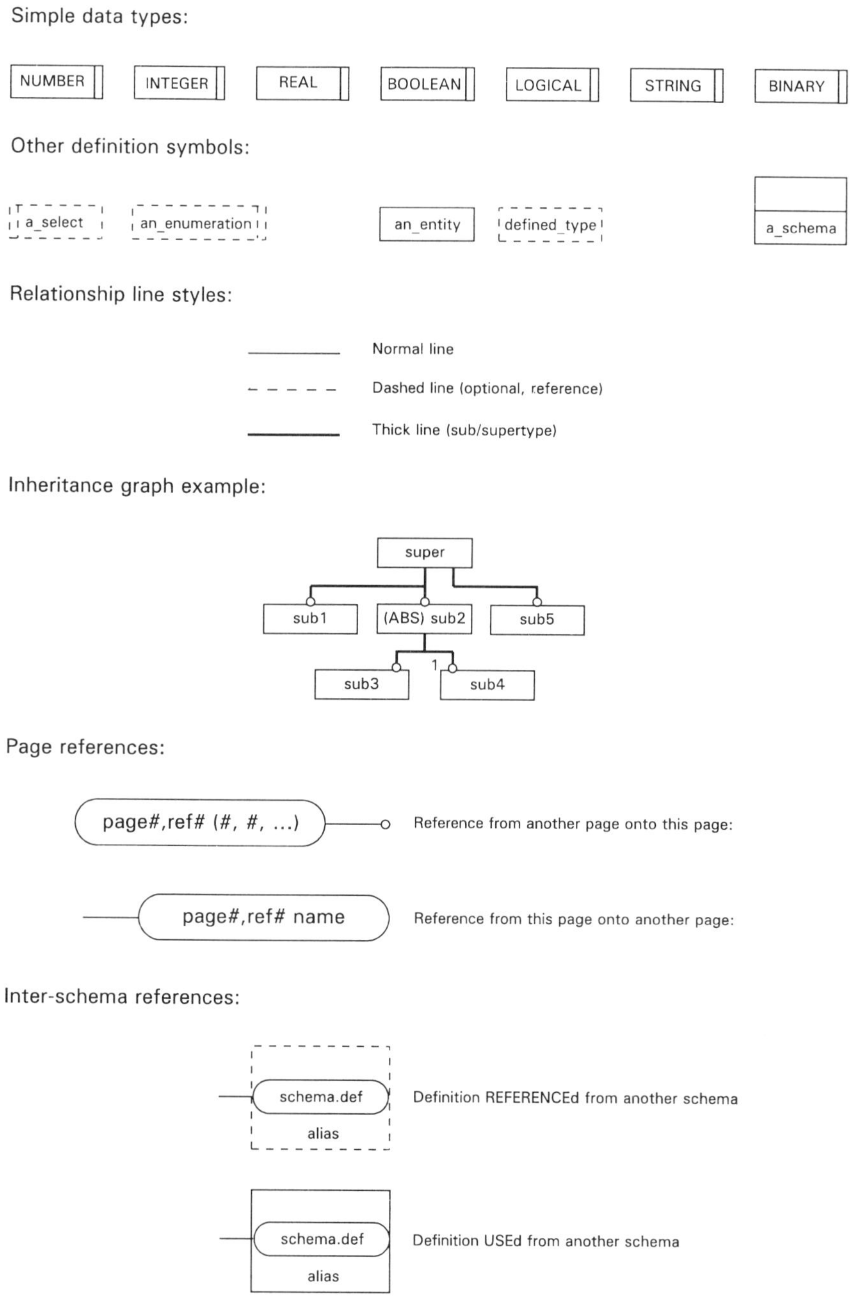

Figure 12.4 EXPRESS-G: key to symbols.

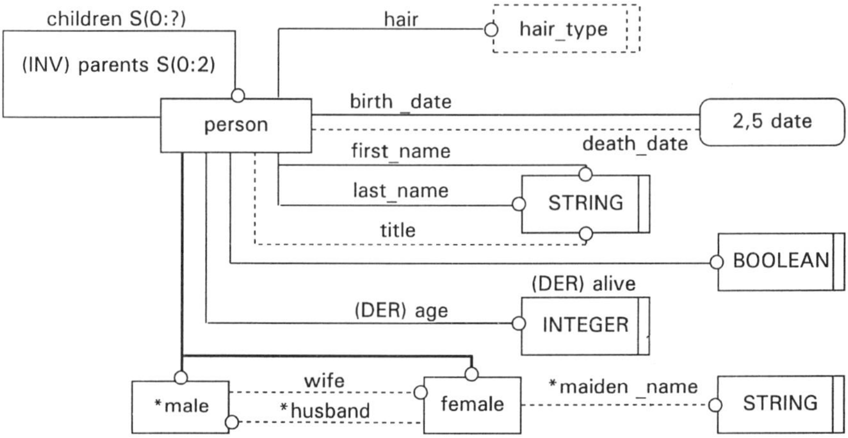

Figure 12.5 Complete entity-level diagram of sample schema shown in Figure 12.3 (page 1 of 2).

Figure 12.6 Complete entity-level diagram of sample schema shown in Figure 12.3 (page 2 of 2).

fundamental importance. It is documented as a normative annex of the EXPRESS document; most of its constructs are shown in Figure 12.4. The schema in Figure 12.3 is presented in its EXPRESS-G form in Figures 12.5 and 12.6, the artificial split into two diagrams showing the EXPRESS-G facility for dividing a large schema across several pages.

Whereas EXPRESS and EXPRESS-G provide means for documenting and defining the structure of information, EXPRESS-I allows instances of that structure to be described; that is, the structure is populated with values. Instances may be displayed at the entity, schema or model level (where a model is a collection of schema instances). Support for defining abstract test cases for conformance testing is also included: it is possible to define a `context`, which may have formal parameters and default values, and a `test case`, which is built from one or more `contexts`. An example is given in Figure 12.14.

12.3 Implementation methods

Part 21 provides a clear-text encoding of the exchange structure: that is, a definition of the format of a physical file. The file itself is sequential and in free format. The syntax of the physical file is defined using Wirth Syntax Notation [174] which means that software can be built quite easily to parse such files.

Table 12.2 Contents of physical file header section.

Entity	Attribution	Explanation
file_description	description	Informal description of contents
	implementation_level	Required implementation level of postprocessor
file_name	name	Name of exchange file
	time_stamp	File creation time and date
	organisation	Organization with which the author is identified
	preprocessor_version	Identification of the software used to create the file and version
	originating_system	The system from which the data originated
	authorisation	Who authorized the sending of the exchange file
file_schema	schema_identifiers	Identifies schema(ta) which specify the instances in the data section of the file (usually the single name of the long form of the application-interpreted model)

A STEP physical file currently comprises two sections – the header and the data sections. Although others may be added in the future, the syntax has been defined in such a way that such extensions will be upwards-compatible. The structure of the header section is shown in Table 12.2: the IGES start and global sections, and the VDA-FS header section contain similar information. The data section contains instances of entities specified in the schema identified in the header section; each entity has a unique identifier.

In order to define how particular EXPRESS constructs appear in the physical file, a mapping is defined in Part 21. Not all EXPRESS constructs need to be mapped to the physical file, as shown in Table 12.3: derived and inverse attributes and defined types are not mapped, for example, whereas explicit attributes are.

Table 12.3 EXPRESS to physical file mapping

EXPRESS primitive	Physical file type	Example
integer	integer	0
real	real	0.0 –0.0E0
string	string	'zero'
binary	binary	%0
array, bag. list, set	list	(0,0)
boolean, enumeration, logical	enumeration	.zero. .true.
entity ;	entity	MALE (attributes)
entity (as attribute)	entity name	#1

Given the EXPRESS schema in Figure 12.3, and the family tree in Figure 12.2, a physical file containing instances of the structures is shown in Figure 12.7. Two points should be noted. First, the file has been laid out with comments and white space: a STEP physical file comprises a string of characters with no carriage returns, although there are requirements set out in Part 21 for such pretty-printing, in case a human needs to read the file. Secondly, because the family tree is necessarily incomplete, the constraints specified in the EXPRESS are violated in the physical file; hence, the file and the processor which produced it are non-conforming to this schema.

```
ISO-10303-21;
HEADER;
FILE_DESCRIPTION(('British royal family details (incomplete)'), '1');
FILE_NAME('ROYAL.STEP', '1992-08-19 T09:00:00',
          ('Jon Owen'), ('The University of Leeds', 'England'),
          'Hand-crafted systems, plc',
          'Generic Almanack 1992, London, England',
          'not approved for release');
FILE_SCHEMA(('ROYAL_EXAMPLE')) ;
ENDSEC;
DATA;
#1 = MALE (
     ('Albert','Frederick','Arthur','George'), /* first name */
     'Windsor', 'George VI',                   /* last name, title */
     (14, 12, 1895), (06, 02, 1952),           /* dates */
     (#2 /* and Margaret Rose */ ), .GREY.,    /* children, hair */
     $ ) ;                                     /* wife (not included) */
#2 = FEMALE ( ('Elizabeth', 'Alexandra', 'Mary'), 'Windsor',
    'Elizabeth II, by the Grace of God, of the United Kingdom of
    Great Britain and Northern Ireland and of her other Realms and
    Territories Queen, Head of the Commonwealth, Defender of the Faith',
    (21, 04, 1926), $, (#3, #8, #6, #7), .GREY., $, $ ) ;
#3 = MALE ( ('Charles', 'Philip', 'Arthur', 'George'), 'Windsor',
    'HRH The Prince of Wales', (14, 11, 1948), $, (#4, #5), .BROWN., #99 )
#8 = FEMALE ( ('Anne', 'Elizabeth', 'Alice', 'Louise'), 'Phillips',
    'HRH The Princess Royal', (15, 08, 1950), $, (), .BROWN., $, $ ) ;
#6 = MALE ( ('Andrew', 'Albert', 'Christian', 'Edward'), 'Windsor',
    'HRH The Duke of York', (19, 02, 1960), $, (), .BROWN., $ ) ;
#7 = MALE ( ('Edward', 'Anthony', 'Richard', 'Louis'), 'Windsor',
    'HRH The Prince Edward', (10, 03, 1964), $, (), .BROWN., $ ) ;
#99 = FEMALE ( ('Diana', 'Frances'), 'Windsor',
    'HRH The Princess of Wales', (01, 07, 1961), $,
    (#4, #5), .FAIR., #3, 'Spencer' ) ;
#4 = MALE ( ('William', 'Arthur', 'Philip', 'Louis'), 'Windsor',
    'HRH Prince William of Wales', (21, 06, 1982), $, (), .FAIR., $ ) ;
#5 = MALE ( ('Henry', 'Charles', 'Albert', 'David'), 'Windsor',
    'HRH Prince Henry of Wales', (15, 09, 1984), $, (), .FAIR., $ ) ;
ENDSEC;
END-ISO-10303-21;
```

Figure 12.7 Example of a STEP physical file.

Given the difficulties of transferring data in machine-readable form with the earlier standards and specifications due to the misuse of magnetic media, Part 21 also includes normative requirements concerning how a STEP physical file is written to magnetic tape or disk.

12.4 Integrated resources

The integrated resources provide a series of general-purpose EXPRESS schemata from which application protocols are constructed. The 40-series parts are truly general-purpose, whereas the 100-series serve a specific – though broad – application or set of applications. Thus, for example, Part 101 provides resources for the draughting-related application protocols, Parts 201 and 202.

Two of the most important parts are those numbered 41 and 42, which respectively provide a framework for product description, and geometric and topological representation.

Part 41 is subdivided into three sections, as shown in Figure 12.8; the schemata are listed in Figure 12.9. The generic product description resources provide a framework which, when combined with the other integrated resource parts, produces a set of integrated product description resources [94]. These, in turn, provide the basis for the development of application protocols. Part 41 thus provides a framework for all types of product, with "hooks" for the representation of shape, material, analysis and kinematics information, as well as the product's presentation.

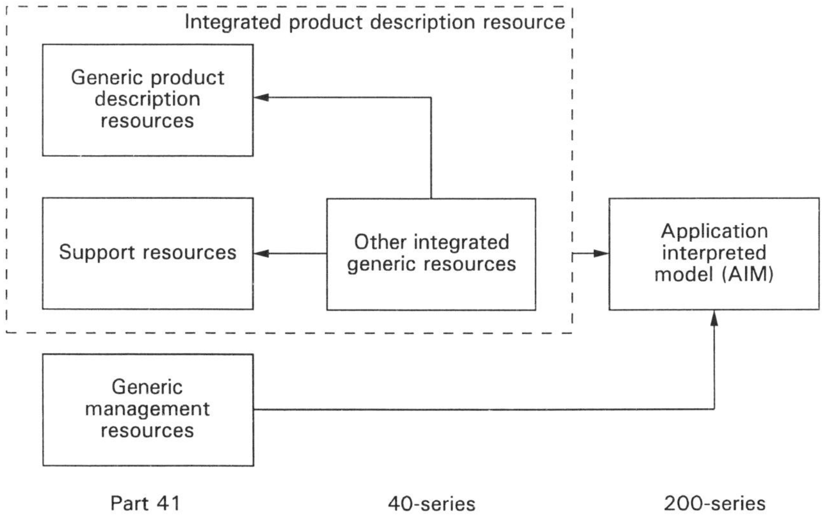

Figure 12.8 STEP architecture for product description and support.

1. generic product description resources
    - application context
    - product definition
    - product property definition
2. generic management resources
    - management resources
3. support resources
    - document
    - action
    - certification
    - approval
    - contract
    - security classification
    - person organization
    - date time
    - group
    - external reference
    - support reference
    - measure
    - effectivity

Figure 12.9 Part 41 schemata.

Part 42 provides the geometry and topology resources for defining the shape of a product. The supertype–subtype structure of each of the three schemata is shown in Figures 12.10, 12.11 and 12.1. This is a level of granularity finer than that presented for Part 41, but serves to show the sorts of geometry and topology that can be defined using Part 42. The geometric representations defined in this part are much better conditioned than those of the earlier specifications. Provision for solid modelling – both constructive solid geometry and boundary representations – has been included from the start.

12.5 Application protocols and abstract test suites

An application protocol comprises four main sections:

(a) Scope.
The scope enables implementors and users to make an initial evaluation of the standard against system capabilities or user requirements. The application protocol project uses the scope to capture and document the industrial requirements that the application protocol addresses. It is documented in one or more of the following:
(i) IDEF0: application activity model;
(ii) IDEF1X, NIAM, EXPRESS-G: planning model;
(iii) English.

(b) Requirements (application reference model).
The application reference model defines a model giving complete details of the information units required to support the activities

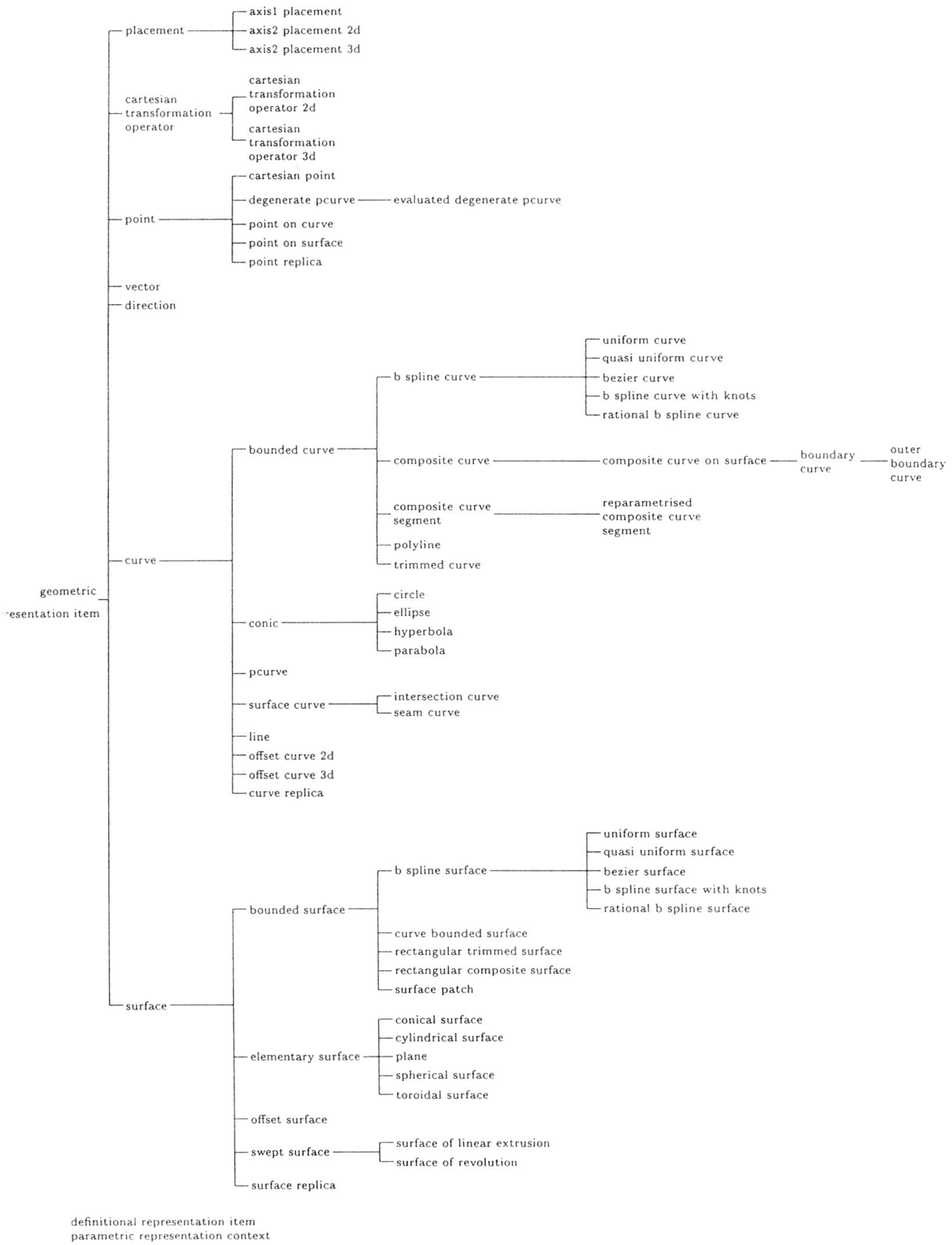

Figure 12.10 Geometry schema.

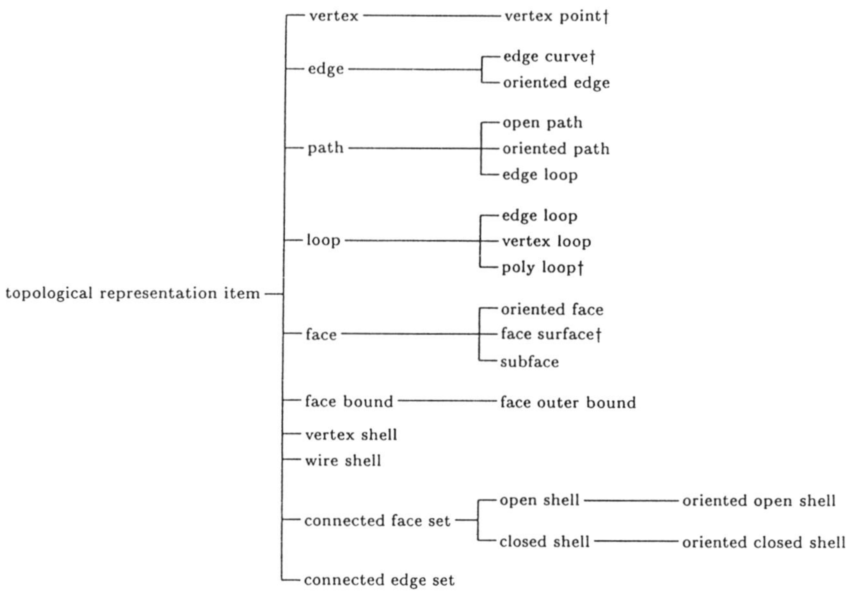

Note: those entities marked with a † are also subtypes of geometric representation item.

Figure 12.11 Topology schema.

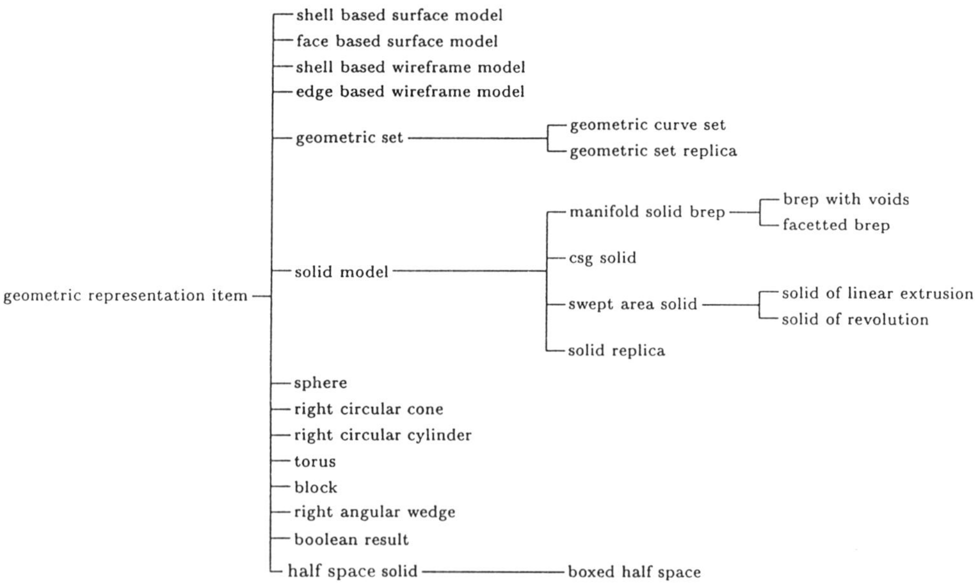

Figure 12.12 Geometric model schema.

identified as lying within the application protocol. It is defined and documented using one of:

(i) NIAM;
(ii) IDEF1X;
(iii) EXPRESS-G.

(c) Application-interpreted model.
Requirements in the application reference model are satisfied in the application-interpreted model by taking the general-purpose (context-independent) integrated resources and interpreting them for use within the application context provided by the application protocol. This interpretation produces an EXPRESS schema, initially in a short form, and a mapping table, showing how each requirement is fulfilled and by which constructs. When the authors of the application protocol are satisfied that the interpretation and mapping table are correct, the long form of the application-interpreted model is produced. This is included in the application protocol document in electronic form only; it is the basis for implementation and is self-contained.

(d) Conformance requirements.
The application reference model is used in conjunction with the long form of the application-interpreted model to develop the detailed test purposes used in conformance testing. Each conforming implementation of an application protocol satisfies the conformance criteria in that individual part (and any normative references made from it).

Some application protocols are themselves substantial documents. As a result, their authors have defined a number of conformance classes within each. For example, Part 203 has six conformance classes and Part 204 has three. These provide options that a vendor may or may not choose to implement, but (as with GKS) complete implementation of a particular conformance class is required for conformance. Each of these conformance classes can itself be regarded as a miniature application protocol. Consequently, users will need to be aware that they need to ask vendors about which conformance classes within an application protocol they actually support. The conformance classes and options are listed explicitly in the PICS proforma, which is documented as a normative annex of each application protocol. This is of fundamental importance for conformance testing.

In practice, application protocols are not constructed from EXPRESS entities, types, functions, and so on. Instead, application-interpreted constructs (AICs) are used, which are collections of the EXPRESS constructs. Thus, an AIC for planar-faceted boundary representations, once defined and documented, can be used as one of the ways of defining shape in Part 204, used within a configuration-controlled design environment in Part

201: the representation of drawings for any phase of the design of a product, including 2D annotation, drawing sheets, revisions, views, grouping of elements, drawing management administration, and the representation of the 2D shape model;

202: similar scope to Part 201, except that associations may be made between dimensions and their host geometry, and that shape is represented by a 3D model (advanced, elementary and facetted B-reps, manifold surfaces with topology, surface geometry without topology, and wireframe geometry both with and without topology);

203: configuration-controlled design information with or without shape (in which shape is represented by wireframe models which have no surface topology, wireframe models with topology, manifold surface models with topology, facetted B-rep and advanced B-rep models);

204: boundary representation solid models (facetted B-rep model, B-rep model with elementary and/or sculptured surfaces) with presentation attributes and product identification and structure information;

205: surface design information (geometrically-bound surfaces, manifold and non-manifold surfaces, all with the option of the presentation of views depicting the shape) with presentation attributes and product identification and structure information;

207: sheet metal part and die set design data, including exchange history, manufacture description, identification of standard parts and purchased items, and design constraints on dies; shape information includes wireframe, surface and solid geometry representations, form features and properties;

209: composite and metallic structural part information required for analysis and related design, including shape (wireframe and surface without topology, wireframe with topology, manifold surfaces with topology, facetted B-reps and advanced B-reps), associated finite element analysis model and analysis results, and material properties;[1]

210: design of electrical printed circuit assemblies, including the identification, description and configuration management of the functional objects in the PCA, reference to analytic models used to define the behaviour, the physical layout of the PCA, and the allocation of requirements to functional objects, physical objects and the physical implementation;

212: electrical aspects required in the design of industrial systems or equipment, including that for power transmission, distribution and generation, electrical machinery, electric light and heat, and control and automation systems;

213: information contained in the numerical control process plans for mechined parts, including references to the resources required to perform the work, the sequence of work instructions, references to documents specified in the process plan, and information to support in-process inspection (such as using gauge blocks or probing information);

214: information which supports the development process of a vehicle, including products of automotive manufacturers and suppliers (car body, power train, chassis and interior) and tools (manufacturing technologies for moulding, casting, bending, stamping, milling, lathing, painting, surface coating, welding and bolting): this includes shape (2D and/or 3D wireframes, surfaces, facetted, elementary and advanced B-reps, compound B-reps and constructive solid geometry), material properties, surface conditions and tolerances, presentation attributes, draughting, kinematics, analysis data for static finite element analysis, identification of standard parts and product structure information;

220: information for the manufacturing process plan for electrical printed circuit boards and assemblies, including process plans, process and factory characterisation for fabrication and assembly, materials used, tolerances and the design of the process itself.

Figure 12.13 The scope of STEP application protocols.

203, and provide shape for associative draughting in Part 202. This enables application-interpreted models to be built more quickly.

The diversity of the areas which the application protocols address – as shown in Table 12.1 – is striking: the emphasis of the second-generation specifications on a single discipline – such as mechanical or electrical – has disappeared. The scope of application protocols which have been balloted is shown in Figure 12.13.

Each application protocol has associated with it an abstract test suite. This latter contains the abstract test cases to be used during conformance

```
CONTEXT tetrashell_instance ;
PARAMETER
  orc : length_measure := 0;                        lx : length_measure := 100;
  ly  : length_measure := 100;                      lz : length_measure := 100;

  origin : cartesian_point := cartesian_point ([orc, orc, orc]);
  p_x    : cartesian_point := cartesian_point ([lx, orc, orc]);
  p_y    : cartesian_point := cartesian_point ([orc, ly, orc]);
  p_z    : cartesian_point := cartesian_point ([orc, orc, lz]);

  neg_x  : direction := direction ([-1, 0, 0]);     neg_y : direction := direction ([0, -1, 0]);
  neg_z  : direction := direction ([0, 0, -1]);
  dslope : direction := direction ([1, 1, 1]);      dperp : direction := direction ([1, -1, 0]);

  loop_x : poly_loop := poly_loop (origin, p_z, p_y) ; loop_y     : poly_loop := poly_loop (origin, p_x, p_z) ;
  loop_z : poly_loop := poly_loop (origin, p_y, p_x) ; loop_slope : poly_loop := poly_loop (p_z, p_x, p_y) ;

  a1 : axis2_placement_3d := axis2_placement_3d (origin, neg_x, neg_y) ;
  a2 : axis2_placement_3d := axis2_placement_3d (origin, neg_y, neg_x) ;
  a3 : axis2_placement_3d := axis2_placement_3d (origin, neg_z, neg_y) ;
  a4 : axis2_placement_3d := axis2_placement_3d (p_x, dslope, dperp) ;

  p1 : plane := plane (a1) ;                        p2 : plane := plane (a2) ;
  p3 : plane := plane (a3) ;                        p4 : plane := plane (a4) ;

  b1 : face_outer_bound := face_outer_bound (loop_x, TRUE) ;
  b2 : face_outer_bound := face_outer_bound (loop_y, TRUE) ;
  b3 : face_outer_bound := face_outer_bound (loop_z, TRUE) ;
  b4 : face_outer_bound := face_outer_bound (loop_slope, TRUE) ;

  fs1 : face_surface := face_surface ([b1], p1, TRUE) ; fs2 : face_surface := face_surface ([b2], p2, TRUE) ;
  fs3 : face_surface := face_surface ([b3], p3, TRUE) ; fs4 : face_surface := face_surface ([b4], p4, TRUE) ;
END_PARAMETER;

SCHEMA_DATA ;
    CONSTANT aic_role : LIST [2:2] OF STRING ==
        ['aic_fac_brep', 'the definition of a B-rep model with planar faces bounded by poly_loops'] ;
    END_CONSTANT ;
    cfs        := connected_face_set {SUBOF(@tri); cfs_faces -> (@fs1, @fs2, @fs3, @fs4]); SUPOF(@tetrashell);}
    tri        := topological_representation_item {SUBOF(@ri); SUPOF(@cfs);} ;
    ri         := representation_item {SUPOF(@tri);} ;
    tetrashell := closed_shell {SUBOF(@cfs); };
END_SCHEMA_DATA;

END_CONTEXT ;
```

The extract from the abstract test case shown below is designed to test the definition of a facetted B-rep containing one (or more) voids. The tetrashell_instance context is used with different parameters to define the outer shell and the void shell. The result is a hollow tetrahedral solid with a large void of a similar shape.

```
TEST_CASE example_204_2; WITH facetted_brep_aic;
    REALIZATION
        CALL tetrashell_instance ; -- uses default values, so no WITH
          IMPORT (shell_object := @tetrashell; aic_info := @aic_role;);
        END_CALL;

        CALL tetrashell_instance ; -- parameters re-set for dimensions (large void)
          IMPORT (hollow1 := @tetrashell; aic_info := @aic_role;);
          WITH (orc := 20; lx := 50; ly := 50; lz := 50;);
        END_CALL;

        void1 := oriented_closed_shell (hollow1, FALSE) ;
        tetra_with_void := facetted_brep (shell_object) || brep_with_voids ([void1]) ;
    END_REALIZATION ;
END_TEST_CASE;
```

Figure 12.14 Example of a STEP abstract text case.

testing of an implementation of the corresponding application protocol. It was decided that such tests were sufficiently important to standardize them. For use in conformance testing, they are converted to executable form, invariably by software, depending upon the implementation method that has been used by the vendor.

An abstract test suite contains:

(a) an administrative header, which includes its date and edition number, and details of the application protocol for which it is the standard abstract test suite;
(b) verdict criteria which apply to groups of test cases;
(c) a list of all test purposes;
(d) abstract test cases, possibly nested into groups.

An abstract test case contains:

(a) test purpose;
(b) test case identifier;
(c) reference to specific parts of the standard(s);
(d) verdict criteria specific to that test case.

It may also contain formal language definitions of the model to exercise the test purpose, written in EXPRESS-I, and statements indicating the construction sequence for the model. This is similar to the contents of the abstract test cases developed by the CTS2 CADCAM project, described in Chapter 10. An extract from an example STEP abstract test case is shown in Figure 12.14.

12.6 Conformance testing methodology and framework

The general procedures for conformance testing have already been described in Chapter 10, so are not repeated here. The IDEF0 (see Appendix C) diagram in Figure 12.15 is similar to that devised for IGES conformance testing, shown in Figure 10.8. The fundamental difference is that most of the controls on the activities are part of the STEP standard itself, rather than being produced by the testing laboratory: the abstract test suite is documented as part of the standard proforma, and also provides the verdict criteria used in analysis; the PICS proforma is part of the application protocol itself; and the abstract test method is provided by Part 34.

Part 31 provides standardized terminology for conformance testing, describes the infrastructure under which conformance testing takes place (see Fig. 10.4), and introduces the basic concepts and framework.

Part 32 enumerates the responsibilities of both the testing laboratory and the client personnel during conformance testing. It is important to reiterate that a client may be any enterprise which provides a system for testing; though this is usually the vendor, it need not necessarily be

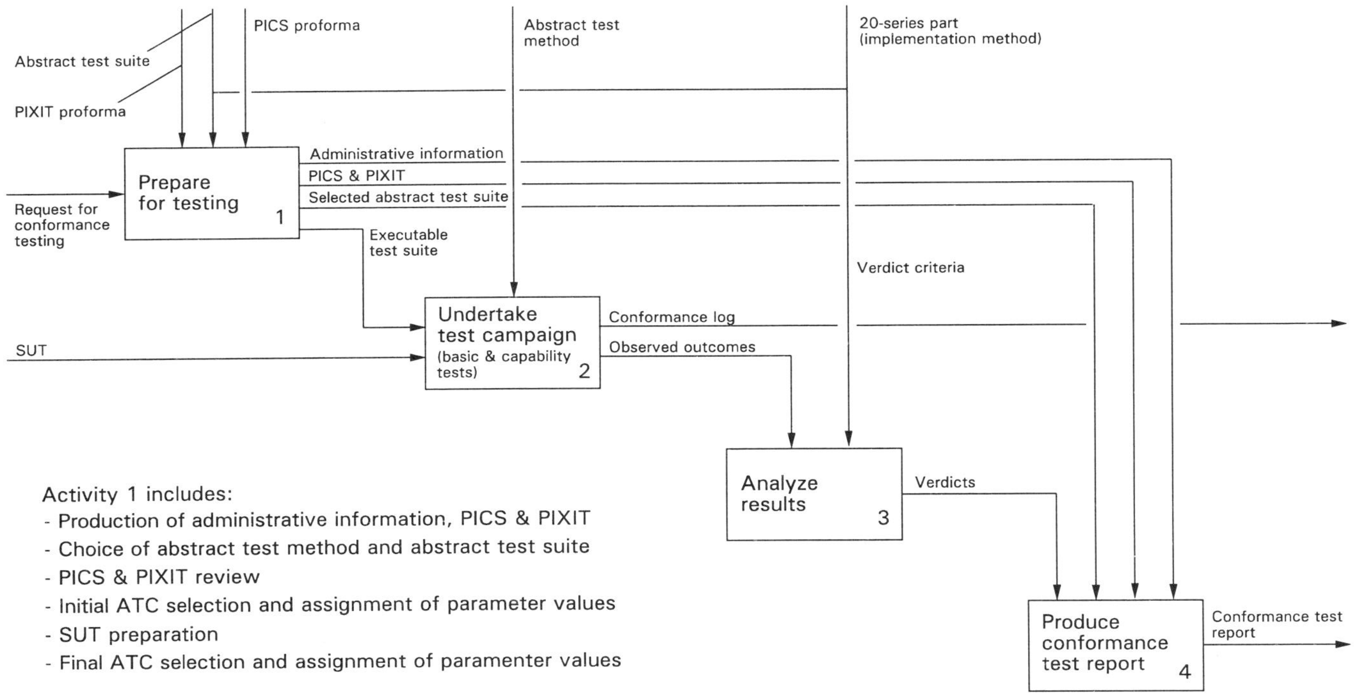

Figure 12.15 Overview of the STEP conformance assessment process.

so. Equally, the testing laboratory is usually independent (see Ch. 8), but may be part of either a vendor or a user enterprise. Provided that it can satisfy the requirements for accreditation, the distinction is not important.

Part 33 describes the structure of an abstract test suite and how it is used during conformance testing, particularly its conversion to executable form.

Part 34 describes the procedures for conformance testing, which vary depending upon the implementation method. Many actions – particularly during the preparation for testing – are common to conformance testing of all implementation methods, but the test campaign will be quite different for a preprocessor physical file and a postprocessor physical file. Reading in the model is trivial when conformance testing a postprocessor physical file ("load the supplied physical file into the CAx system") but potentially very complex for a preprocessor physical file. When further implementation methods are published (such as Part 22, which is described in Ch. 13), Part 34 will need to be expanded or other 30-series parts added.

Concerning the issues of completeness, subsets and conversion, all application protocols so far have specified a number of conformance classes, and that all constructs within the supported conformance class must be supported. Consequently, STEP has adopted the VDA-IS philosophy of "completeness" and "subsets", with the concept of the application protocol itself also providing a method of subsetting the entire standard. No application protocol has yet indicated that an alternative representation of a particular construct is acceptable for conformance, but consideration of a central product model, described in the next chapter, is likely to bring this issue of "conversion", which is well known in conformance testing work, to the attention of more people and to give it a higher priority.

12.7 Conclusion

By the adoption of a three-layer architecture, and the development and documentation of the standard in classes, STEP enabled experts to concentrate on their particular speciality. The concept of application protocols addressed the problems of subsetting and a framework and methodology for conformance testing were developed from the inception of STEP as part of the standard itself. The use of formal description methods and activity and information modelling would reduce the possibilities for ambiguity and hence misinterpretation by a reader. They also provided the means for the standard to be computer-processable as well as human-readable. These techniques and their application are described in the final chapter.

Chapter 13

The way forward

STEP is different from the other standards and specifications described in this text for several reasons.

Its scope is not restricted to a particular industrial sector or discipline, such as automotive design, electrical or aerospace: it is intended to capture the information required at all stages of the life-cycle of a product, from requirements, through conceptual and detailed design, analysis, manufacture, maintenance, ultimately to decommissioning and recycling. Also, the second-generation specifications tended to concentrate upon the capture of geometric information. This was acceptable at the time of development and is still required today; enterprises wish to exchange geometric information for a variety of reasons, including those which initiated this whole area: engineering drawings, surfaces for manufacture and analysis data. However, many enterprises now perceive the geometry as but one aspect of the *product* data they wish to communicate: its shape. There is a need to communicate information concerning its manufacture, its material, its tolerances, its relation to other parts (perhaps in the same family), and the people within an enterprise responsible for the various stages in its development. As a result, STEP is much larger than the initiatives discussed so far. More importantly, it provides a standardized framework for product data, irrespective of the sort of product.

The second difference is that it has benefited from the development of all of the initiatives so far described: the mistakes made in the definition of the early standards and the problems which arose from their industrial use have been learned and (hopefully) rectified. Consequently, it has used a number of formal methods in its development. Activity and information modelling have been used to capture the requirements. The language EXPRESS has been defined as part of the standard itself in order to reduce the ambiguity of natural language definitions. There has been a separation of the various "layers" (application, logical and physical) so that the requirement of an application is not confused with its fulfilment or its realization on a physical medium. An infrastructure and method for conformance testing have been built into the standard from the start, rather than as an afterthought. Standard application protocols have also been defined, meaning that vendors can implement well defined subsets rather than *ad hoc* parts of STEP.

Finally, the effort in defining STEP has been and continues to be international. The work is conducted under the aegis of ISO, meaning that it is not a solution being imposed by one country or a group of countries. Any of these differences would be significant by itself; together, they represent an enormous leap forward in techniques for product data exchange and management.

This text concludes by describing some of the likely developments of STEP in the future. These include extensions to the standard, use of parts of STEP by other standards (such as standard parts), some of the problems which it may face, and how STEP may be adopted by industry.

13.1 Application protocols

The application protocols which have been allocated a part number at the time of writing are shown in Table 12.1. Additional application protocols which are under development are as follows:

- electric/electronic
- product life-cycle
- manufacturing process plans
- polymer testing
- sheet metal
- shipbuilding
- life-cycle management
- product operation
- product procurement
- near net shape processing
- ships electrical systems
- ships HVAC systems
- ships library parts
- ships outfit and furnishings
- constructional steelwork

Many more are likely to be proposed. A summary and status are provided quarterly [65]. The diversity of the areas which the application protocols address has already been noted.

The development process for application protocols has also been documented as part of the STEP project [112]. This is essential for those developing application protocols within the STEP environment, but it is also useful for those outside it. As a result, an enterprise that wishes to exchange product data in a particular area for which a standardized application protocol is not available may develop its own (non-standardized) application protocol. Ultimately, of course, it may become an ISO standard in its own right, but there are several other possibilities, described in Section 13.7, which will benefit the enterprise too.

13.2 Implementation methods

The description of STEP implementation methods in Chapter 12 comprised only that for the physical file. Historically, this was the implementation method used for all previous data exchange standards and specifications: the objective was to exchange product data, and the only way that this could be achieved was to write a physical file to disk, which could then be sent to the receiving postprocessor, usually on magnetic tape, but increasingly on a disk or by a network. Consequently, STEP development in this area concentrated initially upon improving the efficiency of the physical file by building on the experience gained from earlier work. The physical file definition was separated from that of its contents – product data – so that implementation experts could concentrate upon this format. However, this separation enables other implementation methods to be considered for the same structure of product data. The form of the product data is determined by the parent EXPRESS schema; the format defined by Part 21 is a clear-text encoding. Outside STEP, one possible development is an EDIF-compatible file format. Inside STEP, a binary-text encoding, following the example of the computer graphics metafile [27] is possible. Another possibility – that of an application program interface, model access software, or working-form interface, but known within STEP as the SDAI – is currently being developed as Part 22.

Like GKS, the SDAI is a functional specification which is independent of any implementation language. It is realized (in an implementation) in one of several possible programming languages. When combined with a language binding, this provides a callable interface from a program written in that language to data structures which contain information in a form consistent with the EXPRESS schema.

The data is stored in a repository, which may be a file, a working form (in-memory data structures), or a database. The repository may contain several models, the structure of each of which is defined by a single EXPRESS schema. Each schema is self-contained, as the "long form" of the application-interpreted model in an application protocol (see Ch. 12).

The functional specification includes several classes of operations and, again like GKS, uses a finite-state machine approach. It also includes three EXPRESS schemata. Language bindings are being defined for C, Fortran, C++, Ada and Pascal. Further, there are two types of binding: late, which is independent of the EXPRESS schema being implemented, and early, which is dependent upon it. Some of the bindings will be standardized, either as normative annexes to the part or as separate parts in the 20-series.

Because the SDAI allows information to be written and retrieved, there is no concept of a preprocessor or a postprocessor; consequently, conformance testing of an application protocol combined with this implementation method addresses the efficacy of the interface written in a particular high-level programming language. It is possible to do this by compiling programs which test particular aspects and linking them with the (standardized) interface provided by the vendor.

It should be noted that the SDAI does not dictate how the data structures themselves are implemented: only the interface to them is standardized. This leaves vendors free to implement the data structures in the manner they wish, using the advantages that their hardware platform and experience give them. One possible way is to generate data structures which correspond directly to the EXPRESS structures, but established vendors have the legacy of existing systems which precludes – at least partially – this option.

As well as providing an interface to data structures, the SDAI will also allow co-operating processors to access the same central data repository. This would enable a single database to be accessed by a number of sub-contractors. Some would undoubtedly call this concurrent or simultaneous engineering. However, the critical issues of *who* may access particular data *when* have yet to be resolved in STEP. In practice, a design may have to be "signed off" as ready for prototype manufacture, in which case there are strong similarities with an exchange: the differences are that the sign-off can be electronic and no file is written to magnetic tape (although it will almost certainly be archived, having reached this status, for safety).

Four implementation methods were suggested originally for STEP, as described by Smith [141] and Wilson [168]. Of these, the physical file is described in Chapter 12 and the SDAI supports interfaces for model access software and shared databases. The fourth method – intelligent knowledge-based systems – has not yet been addressed in STEP, but the coupling of knowledge-bases and databases has been studied elsewhere [68].

13.3 Conformance and interoperability testing

The infrastructure for STEP conformance testing does not comprise only the 30-series class of parts. These describe – at the appropriate level of abstraction – how conformance testing is to be carried out. However, the test data is also documented: each application protocol has an associated abstract test suite. Within the application protocol itself, the requirements of a conforming implementation are stated. A growing awareness of the importance of this topic is reflected in the division of an application protocol into specific conformance classes and in the inclusion of the PICS proforma indicating the possible options within each application protocol.

Currently, a service for EXPRESS processors is available at CADDETC: the NIST processor was the first to undergo successful testing. It is expected that such services will also become available for particular application protocols, probably based at the national centres. With the issues of accreditation and certification already being addressed, it is likely that the laboratories will offer services for the range of application protocols between them, recognizing one another's results and reports, rather than competing directly. Services may be licensed from one laboratory to another: it is not worth duplicating the effort of establishing a service when the means of testing can be purchased from a partner with which one is collaborating already. There is already interest in Europe in services for those application protocols sponsored originally by the Cadex project (Parts 204, 205 and 206) and in the United States for Part 203.

However, conformance testing is but the first stage in testing an implementation, as noted in Chapter 8. Two conforming implementations may still fail to interoperate because they support different legal options within an application protocol. Conformance testing provides the platform for further types of testing; the next development in this area for STEP will be standardized interoperability testing. This will enable users to obtain an answer to the question "Will this product data in my system transfer to that particular system successfully?" As with conformance testing, there will be no guarantees – testing can only demonstrate errors, not prove correctness – but it will increase the likelihood of success and boost user and vendor confidence further.

13.4 Standard parts

Catalogues of parts are used in all industrial sectors. These may be supplier catalogues, agreed standard collections of parts for a contract, or internal company standards. Current data-exchange specifications cannot cope well with the need to include product data about catalogue parts, possibly by reference. A number of European initiatives, including VDA-PS [156], and the Norwegian CADCOM project, were drawn together under the auspices of CEN/CENELEC. They are now being channelled into a standards initiative in ISO TC184/SC4, which is developing ISO 13584 as well as ISO 10303 (STEP). The former is a separate but related standard for the exchange of parts library information [117]. Like STEP, it is divided into a number of classes:

- conceptual descriptions: (10–19)
- logical resources: (20–29)
- implementation resources: (30–39)
- description methodology: (40–49)
- conformance testing: (50–59)
- view exchange protocol: (101–199)
- standardized content: (200–299)

The parts that have been allocated numbers in ISO 13584

1: Overview and fundamental principles
10: Conceptual model of parts library
20: Generic resources
24: Logical layer of supplier library
26: Identification of library suppliers
31: Programming interface
42: Dictionary methodology
101: Geometrical view exchange protocol by parametric program
201: Geometrical view exchange protocol by conforming specification

The standard separates the representation of part-library information from implementation methods used for exchange, and also separates part-library structure from information about each part or part family. It uses EXPRESS for the definition of the information structures.

A textual definition of a part family may be exchanged using SGML, and a behavioural description of a family of electrical parts by VHDL; ISO 13584 allows references to such descriptions.

13.5 Software development

EXPRESS has been used to specify normative parts of STEP for two reasons: to reduce ambiguity and to facilitate computer processing. Though the former reason has been addressed within this text, the latter has not, until now.

Given that a schema is written in EXPRESS, it can be parsed to check that it is correct both syntactically and semantically (it conforms to the EXPRESS syntax, all references are correctly resolved, all types are visible and matching, and so on). Such a parsing tool does not tell us if the information captured by the schema actually fulfils the requirements that it is supposed to: only human review can do this at present.

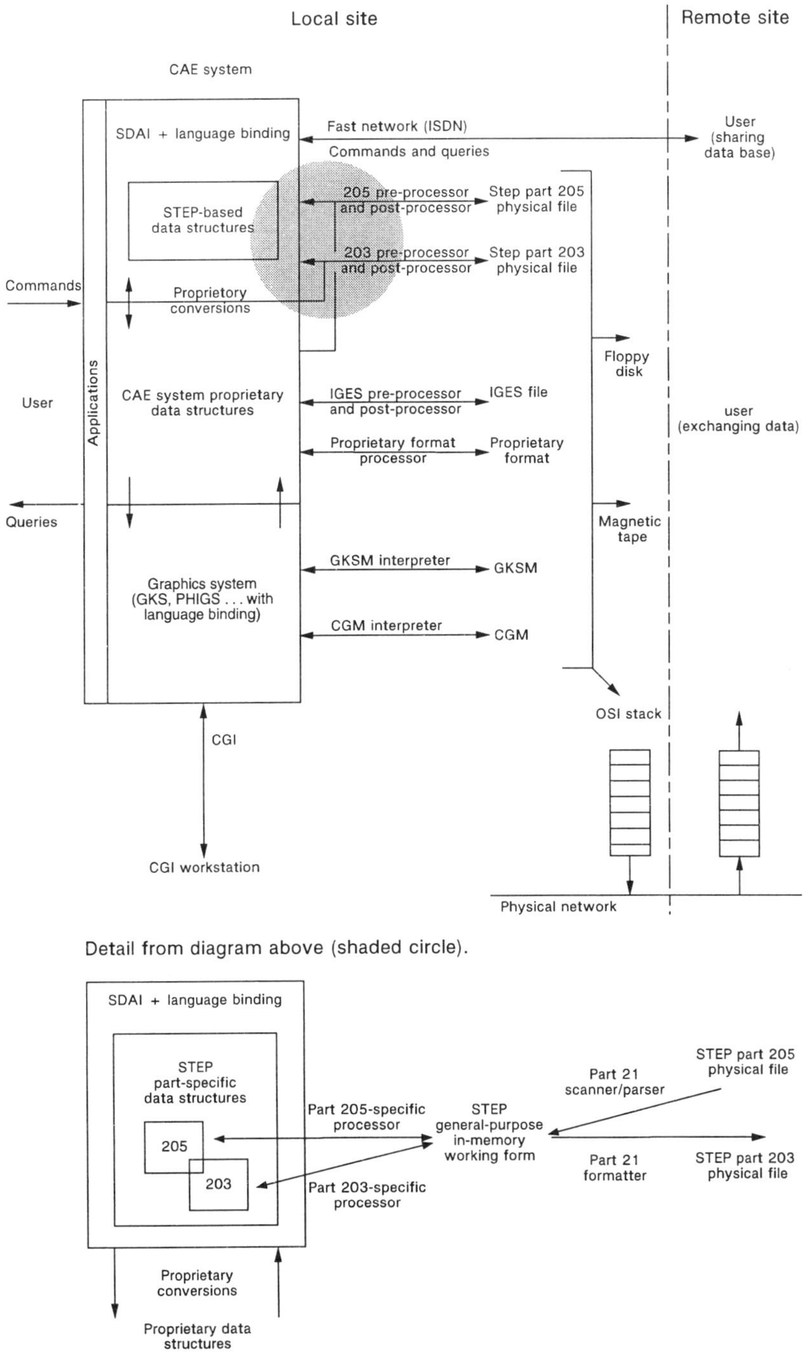

Figure 13.1 A possible STEP software architecture.

However, once the EXPRESS is correct from both points of view, it can become a standard. Such parsing tools may then be expanded to produce more useful results [171]. For example, it is straightforward to add actions into a parser so that as it encounters entities and explicit attributes, it generates the declarations for the corresponding data structures in a particular computer programming language. It is also possible to generate the read and write prescriptions for processing instances of these entities, and the functions which create the data structures. Examples of this have been demonstrated for Fortran, C++, SQL, Ada, dBase and a variety of object-oriented proprietary databases. This makes the writing of a STEP processor easier, because much of the code can be generated automatically from the standardized EXPRESS, which is provided in machine-readable form as part of each application protocol for convenience. The CADEX software architecture, which processes STEP physical files using an SDAI-like interface, is shown in Figure 11.1, and a possible STEP software architecture is shown in Figure 13.1.

13.6 Potential problems

Although STEP has benefited from the expertise gained from other specifications, it would be idle to pretend that all product data exchange problems will disappear when it is adopted. Even when exchange is undertaken using two conformance-tested processors of a well defined standard written using formal methods, problems will arise: as noted in Chapter 8, product data exchange is unlikely to simulate "appliance" or "black-box" technology in the near future, if ever. This section highlights some of the problems that may arise; in practice, they are either longer-term research issues or are solvable by disciplined users and vendors with current technology. Consequently, they should not be interpreted as denigrating the tremendous international initiative that has produced the initial release of STEP and continues to develop it. Rather, it enumerates the next set of issues of be addressed.

The first issue is that of uncontrolled implementations of STEP. Although STEP provides application protocols (possibly containing conformance classes) and a methodology and framework for conformance testing, these are of no value if they are not used. It is up to users to insist on conformance-tested processors for particular application protocols: only their business need will persuade vendors to make the appropriate investment. Equally, conformance testing laboratories need to provide a useful, realistic and cost-effective service, and ensure that both users and vendors understand the implications for a processor that has undergone conformance testing. More education in this area is required: recent conversations with vendors have elicited the phrases

"STEP-conformant", "STEP implementation", "STEP-compliant", "STEP-conversant", and even "STEP-ish" and "STEP-like". Although the first is at least the correct term, it would have been useful to know which of the twenty or so application protocols had actually been implemented using the physical file implementation method. No current system could sensibly implement all of STEP – hence the concept of application protocols; this therefore makes a nonsense of the second term. The other phrases conjure up schemata written in EXPRESS bearing no relation to STEP geometry or draughting, or of STEP models being used where possible and changes made to accommodate the vendor's system, neither of which is remotely concerned with conformance to STEP; perhaps the terms have been coined to confuse a gullible user!

The second issue is that of migration, either from second-generation standards and specifications or from CAx vendors' native formats. This is coupled with the issue of archiving. Naturally, many enterprises will have large amounts of product data archived in these forms, and will only change to using STEP when convinced of the business benefits. First, the amount of space on the physical medium – disk or tape – needs to be less; although the reduction in hardware costs during the development of STEP has ameliorated this somewhat, the amounts of data involved still warrant consideration of the issue. Secondly, the information archived must be complete as well as accurate; unlike exchange, in which the receiving processor and probably its weaknesses are known, the archived data may be retrieved by any postprocessor at any time during the life-cycle of the product. This is particularly long when one considers the survey of a site through to decommissioning. Archiving was one of the original requirements of SET, but seems to have received little specific attention apart from that, perhaps because exchange is a more immediate problem. IGES had limited provision for archiving, with its user-defined associativities and macro language, but it was not common for vendors to implement these facilities; though the structure was preserved, much of the meaning was lost. All STEP application protocols can enable the storage of data specific to an application or individual phase, but none, to date, can enable the storage of *all* of the information concerning a product throughout its life-cycle. Some of the later application protocols may be sufficient to accomplish this. Although migration is not being addressed specifically as part of the standard development, some of the national centres are developing software tools to effect migration to STEP from existing specifications. PROSTEP is developing software tools for migration from VDA-FS and VDA-IS, and the Japanese STEP centre is developing them for IGES.

The third issue concerns the different implementation methods. Each method necessarily includes a mapping from EXPRESS to that method, so that the format of any instance of any EXPRESS model is defined. This means that a single EXPRESS schema drives the appearance of a corresponding physical file or data access interface. The EXPRESS schemata defined in STEP are – quite rightly – not tailored towards a specific implementation method: that would have been contrary to the separation of STEP into classes. However, it is not clear if such generic schemata provide a physical file which is efficient, while providing for a data access interface that is also efficient (as well as for implementation methods yet to be defined). One solution to this possible problem uses a so-called implementation schema, also written in EXPRESS, which takes the general EXPRESS schema and tailors it for a specific implementation method, such as the physical file. For example, a list with an indefinite upper bound could be replaced by a finite array with upper and lower bounds known in advance. This tailoring would have to be undertaken automatically for the concept of standardized models to be retained: the idea of each vendor hand-crafting an EXPRESS schema which had been standardized would not only be inefficient but would lose the benefits gained by standardization. However, until implementations using the physical file and data access interface implementation methods can be investigated and tested for performance, the resolution of this issue cannot progress quickly.

The fourth issue is related to the previous two. Although STEP provides standardized application protocols, and conformance is assessed against an application protocol combined with an implementation method, there is a temptation for vendors to "implement" the integrated resources. There is nothing intrinsically wrong with this, provided claims for "STEP conformance" are not made on this basis. Provided a vendor can fulfil the requirements stated in the application protocol in which the user is interested, no knowledge of the underlying data structures is required. However, all constraints specified in the application protocol EXPRESS schema must be fulfilled in each instance of the data, and entities other than those in the application protocol must not be present. What is not yet clear is if STEP can support a central product model – that is, a single, integrated model that is capable of supporting all aspects of a product throughout its life-cycle [95]. Although Part 214 provides a forum for automotive data, it comprises several distinct disciplines – product structure, shape (geometry and topology), presentation, draughting, robotics and kinematics, finite element analysis, materials, surface conditions, tolerances – most of which are divided into levels. Each of these disciplines could have been documented as a separate application protocol within the automotive sector; these distinct models will need to be made into a single model before the application protocol becomes standardized. A related issue is that, whereas there are experts on geometry, information modelling, analysis, kinematics, materials and so on, there are very few who understand the enabling technologies of integration and interpretation. Given the difficulty of amalgamating models from different areas because of the cross-disciplinary expertise needed, progress in this area will be worth monitoring.

This leads directly to the final issue: that of formal methods. Although both EXPRESS and the physical file format are defined using formal grammars, the integrated resources and application protocols are defined using EXPRESS, and activity and information modelling languages are used to capture requirements, there is still room in STEP for more formalism [170]. In computer graphics, the development of the various standards has culminated in a computer graphics reference model [28], an introduction to which is given by Carson [25]. This has enabled greater understanding of the discipline of computer graphics, particularly the areas that the different standards address. Other work [45] has concentrated upon defining parts of a standard using a formal description language, such as LOTOS [20]. Such an approach has already demonstrated that test purposes can be derived for the standard automatically [118]. This would not only help the conformance testing work, but would enable gaps in the standard to be identified readily, by evaluating such a formal description in the context of a reference model.

Further, the processes of both integration – by which new resource models are integrated with existing ones – and interpretation – by which the application-interpreted model of an application protocol is created from the integrated resources – are informal in nature. Ashworth [9] proposes a language for the latter; addressing this issue will enable the process to become better defined.

13.7 Industrial adoption of STEP

STEP can be adopted by an enterprise, either directly or indirectly, in order to reap the benefits of the technology.

The direct adoption of STEP will mean that STEP physical files are exchanged between enterprises, rather than using IGES, VDA-FS or SET files. However, STEP will not be adopted directly unless it can be shown to work better than the existing generation of product data exchange standards. This means that it has to work more reliably in domains in which it is already possible to exchange data, or that it can be used to exchange data which it is currently not possible to exchange. (In practice, it ought to be able to do both.)

STEP geometry is much better conditioned (having learned from the experience of the existing generation), and its physical file is more efficient, so it is likely to work more reliably than the existing standards in this fundamental domain. It also has a much wider coverage than existing standards, as shown in Table 12.1 and the list in Section 13.1.

Important as the ISO application protocols are, the *method* of constructing them is even more important [112]. By having documented the process, the use of STEP in industry can be applied more widely than simply by adopting the standard directly. We conclude by summarizing from [109] the ways in which an enterprise can adopt STEP.

(a) Use a STEP application protocol directly.
In order for an application protocol to be developed in STEP, an industrial need has to be identified, together with people and funding to develop the corresponding application protocol. With the range of application protocols available already, and those under development, it is likely that one (or a standardized level within one) will satisfy the industrial need of a particular enterprise. As a result, CAx system vendors are likely to have sufficient user pressure to develop implementations of these standard application protocols. Further, conformance testing services will be available at the same time as the standard is published. Such a service includes the provision of the client testing his processor at his own site before presenting it for formal and independent conformance testing by the laboratory. This will help during the development of the implementation as well as for its acceptance testing during procurement, and should result in better quality implementations. At the very least, it will identify and document problems accurately.

(b) Use a restricted, documented subset of a STEP application protocol.
Application protocols have standardized "levels" within them, called conformance classes. However, for particular exchanges, an enterprise may wish to impose additional restrictions on a particular level. By documenting and agreeing these, implementations of standardized application protocols can be used with local restrictions to exchange data between particular enterprises.

(c) Use a national standard application protocol.
Because the application protocol development method has been documented, an industrial need that is useful at a national rather than an international level can be addressed, and documented as a national standard. In the future, it might become a regional or international standard itself.

(d) Use an "industry protocol".
An industrial sector might wish to define its own "industry protocol" without putting it forward for international or national standardization. Given the application protocol development methods, this can be constructed in exactly the same way as the formal standard application protocols, but without the requirement to attend international meetings and go through the international balloting process.

(e) Develop, document and use your own "enterprise protocol" or "organization protocol" [9].

Even more locally, an enterprise is able to document what information it needs. This could be for one particular client, or for all clients with which the enterprise exchanges data. The documentation process can be exactly the same as all of the above.

(f) Use STEP development methods.
An enterprise can still benefit by analysing the information it uses in its activities. Formal activity modelling and information modelling, undertaken as two essential steps in building an application protocol, invariably result in more efficient capture and use of information. Such information can be made available when and where it is needed, and control of that information (or the lack of it) can also be addressed.

(g) Use the understanding gained from developing STEP.
Even if none of the above is adopted, a tremendous amount of expertise and manpower has gone into the development and documentation of STEP. If an enterprise is involved in any aspect of computer-aided engineering, it is worthwhile checking what has been addressed in the STEP arena. It may provoke extra ideas which can be incorporated into the enterprise, or merely save the investigation of an area which clearly is not going to work. The enterprise benefits in either case.

The advantage of using the application protocol development method is not only the greater understanding that results from analysing requirements and so on, but also that an EXPRESS schema is one of the results. A STEP implementation is derived from this schema, partly through automatic software generation: the schema can be parsed, and code generated automatically for data structures, physical-file read/write prescriptions, model access routines, constraint checking, and even conversion algorithms. This means that a CAx vendor will be able to operate on an EXPRESS schema in a STEP application protocol and generate large parts of an implementation of that application protocol. However, the important point is that the program that undertakes the generation does not need to receive EXPRESS schemata that have been standardized: any correct EXPRESS will do.

There is a considerable amount of investment in the existing generation of standards. Consequently, STEP needs to be able to provide migration paths to STEP from standards such as IGES, VDA-FS and SET. This migration can also be effected using the concept of the application protocol, documenting the conversion rules as part of the EXPRESS schema.[7]

If an enterprise documents what information it uses (and when, where, by whom, how and why) it will already be able to become more efficient in the use of that information. If it documents the form of the information in an EXPRESS schema, this can be used as a basis for a software implementation. In these ways, even if STEP is not adopted directly by an enterprise, the benefits it brings to its computer-aided engineering activities are still likely to be large. As a result, data exchange and management can become much more reliable and reproducible, and thereby more efficient and effective.

Appendix A

Abbreviations

This appendix provides acronyms that have been used within this book together with their expansion. Those for national standards bodies are shown in Appendix B. Other acronyms not listed are largely the names of companies.

AEC architectural and engineering construction
AECMA Association Européene des Constructeurs de Matériel Aérospatial
AIC application-interpreted construct
AP application protocol
API application programmers' interface
ARMINES/CPAO Association pour la Recherche et le Développement des Méthodes et Processus Industriels: Centre de Production Assistée par Ordinateur (now IRPL)
ASCII American Standard Code for Information Interchange
ATC abstract test case (used in conformance testing)
BNF Backus–Naur Form (a language for defining syntax)
CAD computer-aided draughting or design (in this book, usually draughting)
CADCAM CAD applied to CAM
CADCOM CAD Communication (Norwegian project)
CADDETC CADCAM Data Exchange Technical Centre (Leeds, UK)
CADEX CAD Exchange (an ESPRIT-funded project)
CAD*I CAD*Interfaces (an ESPRIT-funded project)
CALS Continuous acquistion and life-cycle support (originally: computer-aided acquisition and logistics support)
CAM computer-aided manufacture
CAM-I Computer-Aided Manufacture, Inc.

CAx any computer-aided system used in engineering
CCITT Comité Consultatif International Télegraphique et Télephonique (International Telegraph and Telephone Consultative Committee)
CD Committee Draft (in ISO)
CDC Committee Draft for Comment (in ISO)
CEN Comité Européen de Normalisation (European Committee for Standardization)
CENELEC Comité Européen de Normalisation Electrotechnique
CGI Computer Graphics Interface
CGM Computer Graphics Metafile
CIIN CADCAM Integrated Information Network (Boeing)
CIME Computer Integrated Manufacturing Europe
CO Change Order (in IGES)
CTS Conformance Testing Services (European programme of work)
DBMS database management system
DEKITZ Deutsche Koordinierungsstelle für IT-Normenkonformitätsprüfung und -zertifizierung (German accreditation body)
DIS Draft International Standard (in ISO)
DoD Department of Defense (USA)
DOEDEF Department Of Energy Data Exchange Format (USA)
DP Draft Proposal (in ISO – now termed a CD)
DTI Department of Trade and Industry (UK)
EASE Engineering Applications Support Environment
EBCDIC Extended Binary Coded Decimal Interchange Code
EBNF Extended Backus–Naur Form language for defining syntax
ECITC The European Committee for IT&T Testing and Certification
ECO Edit Change Order (in IGES)
EDI electronic data interchange
EDIF Electronic Design Interchange Format
EDIFACT Electronic Data Interchange Format for Administration, Commerce and Trade
EDM engineering data management
EEC European Economic Community
EFTA European Free Trade Association
EOTC European Organization for Testing and Certification
EPDEN European Product Data Exchange Network
ESP Experimental Solids Project of IPO
ESPRIT European Strategic Programme for Research and Development in Information Technology
ETC executable test case (used in conformance testing)

EU European Union
FE finite element
GKS Graphical Kernel System
GKS-3D Graphical Kernel System for three dimensions
GLATC Graphics and Languages Testing and Certification (agreement group)
GOSET Groupe Opérationel SET (Nanterre, France)
ICAM Integrated Computer-Aided Manufacturing (a US Air Force programme of work)
ICOM input control output mechanism
IDEF integrated definition (originally ICAM definition)
IEC International Electrotechnical Commission
IGES Initial Graphics Exchange Specification
IPC Institute for Interconnecting and Packaging Electronic Circuits
IPIM integrated product information model (in STEP)
IPO IGES/PDES Organization
IRPL Institut de Recherche en Productique et Logistique (Arcueil, France)
IS International Standard (in ISO)
ISO International Organization for Standardization
IVF Institutet för Verkstadteknisk Forskning (Sweden)
KfK Kernforschungszentrum Karlsruhe (Karlsruhe, Germany)
MAP Manufacturing Automation Protocol
NAMAS National Measurement and Accreditation Service
NAPLPS North American Presentation Level Protocol Syntax (a standard for videotext)
NATA National Association of Testing Authorities (Australia)
NAVFAC Naval Facilities Engineering Command
NBS National Bureau of Standards (USA) – now NIST
NC numerical control
NCGA National Computer Graphics Association (USA)
NEDO National Economic Development Office (UK)
NIAM Nijssen's information analysis method
NIST National Institute of Standards and Technology (USA) – previously NBS
NURB non-uniform rational B-spline
NVLAP National Voluntary Laboratory Accreditation Program (USA)
ODA open document architecture (previously office document architecture)
ODETTE Organisation for Data Exchange by TeleTransmission in Europe
OSI Open Systems Interconnection
OSTC Open Systems Testing Consortium (agreement group)

PDDI Product Definition Data Interface (US project)
PDES Product Data Exchange using STEP (originally: Product Data Exchange Specification) – a US programme of work
PDES Inc. PDES Incorporated (a multinational group of companies contributing to the development of STEP)
PDTAG Product Data Technology Advisory Group (Europe)
PHIGS Programmer's Hierarchical Interactive Graphics System
PICS protocol implementation conformance statement
PIXIT protoocol implementation extra information for testing (used in conformance testing)
PMAG Project Management Advisory Group (in ISO)
PSCM Product structure configuration model (in STEP)
PWP printed wiring board
RFC Request For Change (of IGES)
RNE Recherche National d'Essai (French accreditation body)
RP Recommended Practice (for use of IGES)
SADT structured analysis and design technique
SAE Society of Automotive Engineers
SC Sub-committee (in ISO)
SDAI Standard data access interface
SERC Science and Engineering Research Council (now EPSRC)
SET Standard d'Echange et de Transfert (France)
SGML Standard Generalized Markup Language
SI Senter for Industriforskning (Oslo, Norway)
SMMT Society of Motor Manufacturers and Traders (UK)
SPAG Strategic Planning Advisory Group (in ISO)
SPARC Standards Planning And Requirements Committee (ANSI)
STEP Standard for the Exchange of Product Model Data
SUT system under test (used in conformance testing)
TC Technical Committee (in ISO)
TOP Technical and Office Protocol
VDA Verband der Automobilindustrie (Germany)
VDA-FS VDA Flächen Schnittstelle (Germany)
VDA-IS VDA IGES Subsets (Germany)
VDA-PS VDA Programm Schnittstelle (Germany)
VDMA Verband Deutscher Maschinen- und Anlagenbau e.V. (Germany)
VHDL VHSIC Hardware Description Language
VHSIC very high speed integrated circuit
WD Working Draft (in ISO)
WG Working Group (in ISO)
WSN Wirth Syntax Notation (a language for defining syntax)
XBF experimental boundary file
XSET SET exchange software

Appendix B

Standards bodies

Australia
SAA
Standards Australia
PO Box 458 AUS – North Sydney
NSW 2059

Phone: + 61 2 963 4111
Fax: + 61 2 959 3896

Belgium
IBN
Institut Belge de Normalisation
Avenue de la Brabançonne 29
B-1040 Bruxelles

Phone: + 32 2 734 92 05
Fax: + 32 2 733 42 64

Brazil
ABNT
Associação Brasileira de Normas Técnicas
Av 13 de Maio, *n*°13–28° andar
Ciaixa Postal 1680
CEP: 20.003 – Rio de Janeiro – RJ

Phone: + 55 21 210 31 22

Canada
SCC
Standards Council of Canada
350 Sparks Street, Suite 1200
CDN – Ottawa, Ontario K1P 6N7

Phone: + 1 613 238 3222
Fax: + 1 613 995 4564

CEN
CEN
European Committee for Standardization
Rue de Stassart, 36
B-1050 Bruxelles
Belgium

Phone: + 32 2 519 68 11
Fax: + 32 2 519 68 19

France
AFNOR
Association Française de Normalisation
Division Informatique
Tour Europe
Cedex 7
F-92049 Paris la Défense

Phone: + 33 1 42 91 55 55
Fax: + 33 1 42 91 56 56

Germany
DIN
Deutsches Institut für Normung
Burggrafenstraße 6
Postfach 11 07
Berlin

Phone: + 49 30 26 011
Fax: + 49 30 260 12 31

Hungary
MSZH
Magyar Szabványügyi Hivatal
Pf 24
H-1450 Budapest 9

Phone: + 36 1 118 30 11
Fax: + 36 1 118 51 25

ISO
ISO
International Organization for Standardization
Case Postale 56
CH-1211 Genève 20
Switzerland

Phone: + 41 22 749 01 11
Fax: + 41 22 733 34 30

Italy
UNI
Ente Nazionale Italiano di Unificazione
Piazza Armando Diaz 2
I-20123 Milano

Phone: + 39 2 72 00 11 41
Fax: + 39 2 869 01 20

Japan
JISC
Japanese Industrial Standards Committee
Standards Department
Agency of Industrial Science and Technology
Ministry of International Trade and Industry
1-3-1 Kasumigaseki Chiyoda-ku
J-Tokyo 100

Phone: + 81 3 501 92 95 or 96
Fax: + 81 3 580 14 18

Netherlands
NNI
Nederlands Normalisatie Instituut
Kalfjeslaan 2
Postfach 50 59
NL-2600 GB Delft

Phone: + 31 15 69 03 90
Fax: + 31 15 69 01 90

Norway
NSF
Norges Standardiseringsforbund
Postboks 7020
Homansbyen
N-0306 Oslo 3

Phone: + 47 22 46 60 94
Fax: + 47 22 46 44 57

Poland
PKNMiJ
Polish Committee for Standardization, Measures and Quality Control
Ul. Elektoralna 2
PL-00-139 Warsaw

Phone: + 48 22 20 54 34
Fax: + 48 22 20 66 46

Sweden
SIS
Standardiseringskommissionen i Sverige
Box 3295 Tégnergatan 11
S-10366 Stockholm 6

Phone: + 46 8 613 52 00
Fax: + 46 8 11 70 35

Switzerland
SNV
Swiss Association for Standardisation
Kirchenweg 4
Postfach
CH-8032 Zürich

Phone: + 41 1 384 47 47
Fax: + 41 1 384 47 74

UK
BSI
British Standards Institution
389 Chiswick High Road
London
W4 4AL

Phone: + 44 181 996 9000
Fax: + 44 181 996 7400

USA
ANSI
American National Standards Institute
1430 Broadway
New York
NY 10018

Phone: + 1 212 354 3300
Fax: + 1 212 302 1286

Appendix C

Enabling technologies

In practice, many of the techniques used in the development of STEP were also used in the definition of the second generation standards and specifications for product data exchange, in research projects, and in related standards. Rather than describe them as part of the initiative in which they were used, they are gathered into this appendix for convenience.

C.1 The ICAM programme

The ICAM programme was funded by the US Air Force to provide structured methods for applying computer technology to manufacturing. Three methods were documented initially, which would enable models of manufacturing systems or environments to be defined:

- IDEF0: function model (structured representation of the functions and of the information and objects which interrelate these functions) [71];
- IDEF1: information model (structure of information needed to support functions) [72];
- IDEF2: dynamics model (time-varying behaviour of the functions, information and resources) [74].

Since that time, many more such definition languages have been developed and documented, including IDEF3 [92] and IDEF4 [93]. An overview of all of these methods is given by Rupp et al. [129].

C.2 Activity modelling

IDEF0 [71] is a method for modelling activities. It was originally developed by Douglas Ross in the late 1960s, and is the public-domain version of Softech's SADT (structured analysis and design technique) [91, 127, 128]. The result of applying IDEF0 is a model, which has a specific viewpoint, and which comprises diagrams, text and a glossary, cross-referenced to each other. The major component is the diagrams, which provide top-down decomposition with successive introduction of detail. Thus, an IDEF0 model is a hierarchically organized set of diagrams, with a single master diagram at the root and a number of subdiagrams forming a series of levels.

On the diagrams, activities are represented by boxes, which are connected by arrows. Each activity may be decomposed into a number of subactivities, thus producing a hierarchically organized set of diagrams. Each activity may have inputs, controls, outputs and mechanisms (denoted by ICOM codes), as shown in Figure C.1.

An example of a master "root" diagram is shown in Figure 10.7, with its initial breakdown in Figure 10.8. The structure of the rest of the model is described in Figure 10.9.

The decomposition may take place to any depth that is useful, and may vary in different parts of the model. However detailed the modelling is, the method results in the definition of which information is required by and produced by each subactivity. It also enables each box to be used as the basis for writing a procedure that is tractable and has a defined interface with the rest of the model. In some cases, the activity is undertaken by a software executive, for which the inputs, outputs, controls and mechanisms provide detail for its requirements specification.

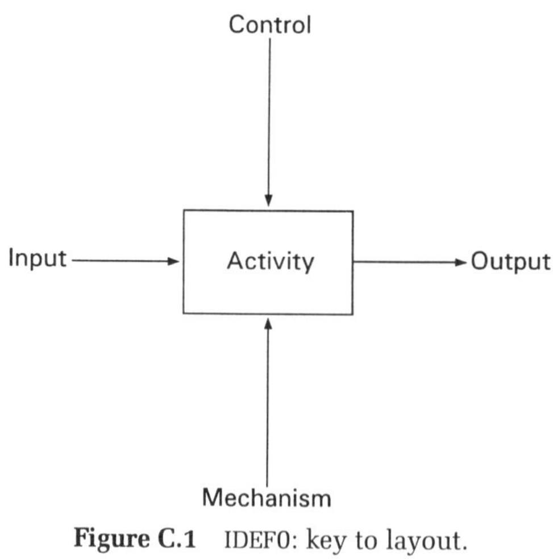

Figure C.1 IDEF0: key to layout.

C.3 Information modelling

IDEF1X [73] and NIAM [157] are both used in STEP for information modelling [167]. Once an activity model has been created, a complete information model can be created, in order to answer the following questions:

(a) What is the data?
(b) What form does it take?
(c) On what does it depend?
(d) How is it identified?

IDEF1X is an extension of the original IDEF1, and allows a user to model entities, attributes and relationships. Once the entities and the relationships between them have been identified, their attributes can be defined. Each instance of an entity has a unique value for its key attribute (or attributes); such key attributes migrate to the children of the entity, in much the same way as an EXPRESS supertype's attributes and constraints. Subtype and supertype mechanisms are also available in the language.

NIAM enables binary relations between entities to be specified: an entity A has a particular relation to entity B, but B also has a relation with A. NIAM provides several types of constraints, and also includes a subtype and supertype mechanism.

More details on IDEF1X are given by Williams [161] and a comparison of IDEF1X, NIAM and EXPRESS (as it was at that time) is provided by Williams [162].

C.4 Wirth syntax notation

WSN (Wirth syntax notation) [174] is a syntax description language which is used to define programming and other languages. This metalanguage can be used to define its own syntax, as shown in Figure C.2.

```
syntax =       { production } .
production =   identifier "=" expression "." .
expression =   term { "|" term } .
term =         factor { factor } .
factor =       identifier
               | literal
               | "{" expression "}"
               | "[" expression "]"
               | "(" expression ")" .
identifier =   letter { letter } .
literal =      """" character { character } """"
```

Figure C.2 Wirth Syntax Notation.

The vertical bar separates two alternatives; the curly brackets denote repetition – that is, their contents may be omitted or appear once or many times – and square brackets denote optionality: the contents may be included or omitted. Parentheses serve merely to group. In order to represent a double quote (") inside a literal, it is repeated.

A further example, that of the definition of the syntax of a real number, is shown in Figure C.3. It is both concise and unambiguous.

```
real =
           [ sign ]
           digit { digit } "."
           { digit }
           [ ( "E" | "D" ) [ sign ] digit { digit } ]
```

`0.`	valid
`-0.`	valid
`0.00`	valid
`0.00E0`	valid
`0.00D-00`	valid
`.0`	invalid: no integral digit
`0.00E*0`	invalid: illegal sign in exponent
`0.0e0+E0`	invalid: illegal lower-case letter in fractional part

Figure C.3 Example of Wirth Syntax Notation.

Appendix D

IGES coverage

This appendix presents the version of IGES in which each entity and form was introduced.

In all cases, the latest name for a given entity or form is reproduced. It also documents the coverage of each class of MIL-D-28000A, the VDA-IS subsets and the CADDETC IGES Processor Conformance Testing Service (marked with a ✓). (IGES versions are given as Arabic numerals, MIL-D-28000A classes as Roman numerals, and VDA-IS subsets as one of: g1, g2, g3, b1, b2, af1, af2.)

Deprecated entities are marked with a †; an additional table is provided at the end giving a summary of such entities and their recommended replacements.

Geometry entities

Entity	Form	Description	Version introduced in	Coverage					
0		null	4.0			I	II	III	IV
100		circular arc	1.0	✓	g1	I	II	III	IV
102		composite curve	1.0	✓	g1	I	II	III	IV
104		conic arc							
	0	determined by parameters	1.0	✓	g1	I	II	III	IV
	1	elliptic	1.0	✓	g1	I	II	III	IV
	2	hyperbolic	1.0	✓	g1	I	II	III	IV
	3	parabolic	1.0	✓	g1	I	II	III	IV
106		copious data							
	1	data pairs in a plane	1.0		g1				IV
	2	data triples in a plane	1.0		g2				IV
	3	data sextuples in a plane	1.0		g3				
	11	2D linear path	1.0	✓	g1	I	II	III	IV
	12	3D linear path	1.0		g2		II	III	IV
	13	piecewise linear string	1.0		g3				
	20	centreline through points	1.0				II	III	IV
	21	centreline through circle centres	1.0				II	III	IV

Entity	Form	Description	Version introduced in	Coverage					
	31	section: solid	1.0		b1		II		
	32	section: solid pairs	1.0				II		
	33	section: solid + dash	1.0				II		
	34	section: solid quads	1.0				II		
	35	section: solid + dash + solid	1.0				II		
	36	section: dash	1.0				II		
	37	section: solid + perpendicular solid	1.0				II		
	38	section: solid + perpendicular dash	1.0				II		
	40	witness line	1.0	✓	b1		II	III	IV
	63	simple closed planar curve	2.0	✓		I	II	III	IV
108		plane							
	−1	bounded void	1.0					III	IV
	0	unbounded	4.0	✓	g3b2		II	III	IV
	1	bounded	1.0		g3		II	III	IV
110		line	1.0	✓	g1	I	II	III	IV
112		parametric spline curve	1.0	✓	g1	I	II	III	
114		parametric spline surface	1.0	✓	g3		II		
116		point	1.0	✓	g1		II	III	IV
118		ruled surface							
	0	equal relative arc length	1.0				II		
	1	equal relative parametric values	2.0		g3		II		IV
120		surface of revolution	1.0		g3		II		IV
122		tabulated cylinder	1.0		g3		II		IV
123		direction	5.1						
124		transformation matrix							
	0	orthonormal rotate/translate	1.0	✓	g1	I	II	III	IV
	1	orthonormal rotate-reflect/translate	3.0				II	III	IV
	10	cartesian co-ordinates	2.0						
	11	cylindrical co-ordinates	2.0						
	12	spherical co-ordinates	2.0						
125		flash							
	0	defined by referenced entity	2.0					III	
	1	circular	2.0					III	
	2	rectangle	2.0					III	
	3	doughnut	2.0					III	
	4	canoe	2.0					III	
126		rational B-spline curve							
	0	determined by parameters	2.0	✓	af1	I	II	III	IV
	1	line	2.0		af1	I	II	III	IV
	2	circular arc	2.0		af1	I	II	III	IV
	3	elliptical arc	2.0		af1	I	II	III	IV
	4	parabolic arc	2.0		af1	I	II	III	IV
	5	hyperbolic arc	2.0		af1	I	II	III	IV
128		rational B-spline surface							
	0	determined by parameters	2.0	✓	af1		II		IV
	1	plane	2.0		af1		II		IV
	2	right circular cylinder	2.0		af1		II		IV
	3	cone	2.0		af1		II		IV
	4	sphere	2.0		af1		II		IV
	5	torus	2.0		af1		II		IV
	6	surface of revolution	2.0		af1		II		IV

Geometry entities

Entity	Form	Description	Version introduced in	Coverage			
	7	tabulated cylinder	2.0	af1	II		IV
	8	ruled surface	2.0	af1	II		IV
	9	general quadric surface	2.0	af1	II		IV
130		offset curve	3.0		II		IV
132		connect point	3.0			III	
134		node	2.0				
136		finite element	2.0				
138		nodal displacement and rotation	3.0				
140		offset surface	3.0		II		IV
141		boundary	5.0				
142		curve on a parametric surface	3.0	af1	II		IV
143		bounded surface	5.0				
144		trimmed parametric surface	3.0	af1	II		IV
146		FEM nodal results	4.0				
148		FEM element results	4.0				
150		block	4.0				
152		right angular wedge	4.0				
154		right circular cylinder	4.0				
156		right circular cone frustrum	4.0				
158		sphere	4.0				
160		torus	4.0				
162		solid of revolution					
	0	closed to axis	4.0				
	1	closed to itself	4.0				
164		solid of linear extrusion	4.0				
168		ellipsoid	4.0				
180		boolean tree	4.0				
182		selected component	5.0				
184		solid assembly	4.0				
186		manifold solid B-rep object	5.1				
190		plane surface					
	0	unparameterized surface	5.1				
	1	parameterized surface	5.1				
192		right circular cylindrical surface					
	0	unparameterized surface	5.1				
	1	parameterized surface	5.1				
194		right circular conical surface					
	0	unparameterized surface	5.1				
	1	parameterized surface	5.1				
196		spherical surface					
	0	unparameterized surface	5.1				
	1	parameterized surface	5.1				
198		toroidal surface					
	0	unparameterized surface	5.1				
	1	parameterized surface	5.1				

Notes

1. There are six *types* (as opposed to *forms*) for entities 112 and 114: 1 (linear), 2 (quadratic), 3 (cubic), 4 (Wilson–Fowler), 5 (modified Wilson–Fowler) and 6 (B-spline).
2. In VDA-IS for entity 114, only type 3 is allowed.
3. There are thirty-five forms for entities 146 and 148; they have not been included in the main tables, but are reproduced in Table D.1.

Annotation entities

Entity	Form	Description	Version introduced in	Coverage					
202		angular dimension	1.0	✓	b1		II	III	IV
204		curve dimension	5.0						
206		diameter dimension	1.0	✓	b1		II	III	IV
208		flag note	1.0						
210		general label	1.0	✓	b1		II	III	IV
212		general note							
	0	simple note (default)	1.0	✓	b1	I	II	III	IV
	1	dual stack	3.0		b1	I	II	III	IV
	2	imbedded font change	3.0		b1	I	II	III	IV
	3	superscript	3.0		b1	I	II	III	IV
	4	subscript	3.0		b1	I	II	III	IV
	5	super-/subscript	3.0		b1	I	II	III	IV
	6	multiple stack–left justified	3.0		b1	I	II	III	IV
	7	multiple stack–centre justified	3.0		b1	I	II	III	IV
	8	multiple stack–right justified	3.0		b1	I	II	III	IV
	100	simple fraction	3.0		b1	I	II	III	IV
	101	dual stack fraction	3.0		b1	I	II	III	IV
	102	imbedded font change–double fraction	3.0		b1	I	II	III	IV
	105	super-/subscript fraction	3.0		b1	I	II	III	IV
213		new general note	5.0						
214		leader (arrow)							
	1	wedge	1.0	✓	b1		II	III	IV
	2	triangle	1.0	✓			II	III	IV
	3	filled triangle	1.0	✓			II	III	IV
	4	no arrowhead	1.0	✓	b1		II	III	
	5	circle	1.0	✓			II	III	
	6	filled circle	1.0	✓			II	III	
	7	rectangle	1.0	✓			II	III	
	8	filled rectangle	1.0	✓			II	III	
	9	slash	1.0	✓	b1		II	III	
	10	integral sign	1.0	✓	b1		II	III	
	11	open triangle	3.0	✓			II	III	IV
	12	dimension origin	5.0						
216		linear dimension							
	0	undetermined form	1.0	✓	b1		II	III	IV
	1	diameter form	5.0						
	2	radius form	5.0						
218		ordinate dimension							
	0	witness line or leader	1.0	✓	b1		II	III	IV
	1	witness line and leader	5.0						
220		point dimension	1.0	✓	b1		II	III	IV
222		radius dimension							
	0	single-leader format	1.0	✓	b1		II	III	IV
	1	multiple-leader format	4.0				II	III	IV
228		general symbol							
	0	general symbol	3.0				II	III	IV
	1	datum feature	4.0				II	III	IV
	2	datum target	4.0				II	III	IV
	3	feature control frame	4.0				II	III	IV
	5001..9999	implementor defined	5.0						
230		sectioned area							
	0	standard cross-hatching	3.0	✓		I	II	III	
	1	inverted cross-hatching	5.0						

Definition entities

Entity	Form	Description	Version introduced in	Coverage					
302		associativity definition	1.0		af2			III	
304		line font definition							
	1	subfigure pointer	1.0				II		
	2	repeating pattern	1.0				II		
306		macro definition	1.0						
308		subfigure definition	1.0	✓		I	II	III	
310		text font definition	1.0						
312		text display template							
	0	absolute co-ordinates	3.0					III	
	1	incremental co-ordinates	3.0					III	
314		colour definition	3.0				II		
316		units data	5.0						
320		network subfigure definition	3.0					III	
322		attribute table definition							
	0	one-to-many	4.0					III	
	1	one-to-one	4.0					III	
	2	one-to-one with text display template	4.0					III	

Structure entities

Entity	Form	Description	Version introduced in	Coverage					
402		associativity instance							
	1	unordered group with backpointers	1.0		g1			III	
	2	external logical reference file index †	3.0						
	3	views visible	1.0		b2		II		IV
	4	views visible, colour, line weight	1.0				II		IV
	5	entity label display	1.0						
	6	view list †	1.0						
	7	unordered group without backpointers	1.0	✓	g1		II	III	IV
	8	signal string †	2.0						
	9	single parent	2.0						IV
	10	text node †	2.0						
	11	connect node †	2.0						
	12	external reference file index	3.0					III	
	13	dimensioned geometry †	3.0						
	14	ordered group with backpointers	3.0						
	15	ordered group without backpointers	3.0						
	16	planar group	3.0						
	18	flow	3.0					III	
	19	segmented views visible	5.0						
	20	piping flow	5.1						
	21	dimensioned geometry	5.1						
	5001..9999	implementor defined (302 required)	1.0		af2			III	
404		drawing							
	0	translate and scale	1.0	✓	b2	I	II	III	IV

Structure entities

Entity	Form	Description	Version introduced in	Coverage					
	1	translate, scale and rotate	5.0						
406		property							
	1	definition levels	1.0				II	III	
	2	region restriction	2.0					III	
	3	level function	2.0				II	III	
	4	region fill †	2.0						
	5	line widening	2.0				II	III	
	6	drilled hole	2.0					III	
	7	reference designator	2.0						
	8	pin number	2.0						
	9	part number	2.0						
	10	hierarchy	2.0						
	11	tabular data	3.0						
	12	external reference file list	3.0					III	
	13	nominal size	3.0						
	14	flow line specification	3.0						
	15	name	3.0			I	II	III	IV
	16	drawing size	3.0		b2	I	II	III	IV
	17	drawing units	3.0			I	II	III	IV
	18	inter-character spacing	4.0			I			
	19	line font	5.0						
	20	highlight	5.0						
	21	pick	5.0						
	22	uniform rectangular grid	5.0						
	23	assocativity group type	5.0						
	24	level to PWB layer map	5.0						
	25	PWB artwork stackup	5.0						
	26	PWB drilled hole	5.0						
	27	generic data	5.1						
	28	dimension units	5.1						
	29	dimension tolerance	5.1						
	30	dimension display data	5.1						
	31	basic dimension	5.1						
	5001..9999	implementor defined	1.0						
408		singular subfigure instance	1.0	✓		I	II	III	
410		view							
	0	orthographic parallel	1.0	✓	b2	I	II	III	IV
	1	perspective	5.0						
412		rectangular array subfigure instance	2.0			I		III	
414		circular array subfigure instance	2.0			I		III	
416		external reference							
	0	one definition from library file	3.0					III	
	1	full referenced library file	3.0					III	
	2	external logical references	3.0					III	
	3	native system subfigure	5.0						
	4	native library subfigure	5.1						
418		nodal load and constraint	3.0						
420		network subfigure instance	3.0					III	

Structure entities

Entity	Form	Description	Version introduced in	Coverage
422		attribute table instance		
	0	single row	4.0	III
	1	multiple row	4.0	III
430		solid instance	4.0	
502		vertex	5.1	
504		edge	5.1	
508		loop	5.1	
508		edge	5.1	
510		face	5.1	
514		shell	5.1	
600..699		implementor-defined macro instance	1.0	
5001..9999		implementor-defined macro instance	3.0	
10000..99999		implementor-defined macro instance	3.0	

Deprecated entities

Deprecated entity		Recommended replacement		IGES version
402/2	external logical reference file index	402/12	external reference file index	5.0
402/6	view list	402/3	views visible	3.0
402/8	signal string	402/18	flow associativity	3.0
402/10	text code	312	text display template	3.0
402/11	connect node	132	connect point	3.0
402/13	dimensioned geometry	402/21	dimensioned geometry	5.1
406/4	region fill property	230	sectioned area	4.0

Table D.1 Forms for IGES entities 146 and 148

Form	Description
0	unknown/miscellaneous
1	temperature
2	pressure
3	total displacement
4	total displacement and rotation
5	velocity
6	velocity gradient
7	acceleration
8	flux
9	elemental force
10	strain energy
11	strain energy density
12	reaction force
13	kinetic energy
14	kinetic energy density
15	hydrostatic pressure
16	coefficient of pressure
17	symmetric 2-dimensional elastic stress tensor
18	symmetric 2-dimensional total stress tensor
19	symmetric 2-dimensional elastic strain tensor
20	symmetric 2-dimensional plastic strain tensor
21	symmetric 2-dimensional total strain tensor
22	symmetric 2-dimensional thermal strain tensor
23	symmetric 3-dimensional elastic stress tensor
24	symmetric 3-dimensional total stress tensor
25	symmetric 3-dimensional elastic strain tensor
26	symmetric 3-dimensional plastic strain tensor
27	symmetric 3-dimensional total strain tensor
28	symmetric 3-dimensional thermal strain tensor
29	general elastic stress tensor
30	general total stress tensor
31	general elastic strain tensor
32	general plastic strain tensor
33	general total strain tensor
34	general thermal strain tensor

Appendix E

IGES entity prescriptions

This appendix presents the parameter data for IGES entities, in a notation for which the key is given in Figure E.1.

Complete prescriptions have not been included for the entities 0, 306, 402 implementor-defined forms, 422, 600 and above (implementor-defined) because these require information other than that included in the parameter data section of the entity itself.

I	integer	Nchar	integer used as a counter
R	real	L	logical (0 or 1)
P	pointer to a directory entry record	S	Hollerith string
A	single value (I, R, L, P or S)	2R	point in 2D
3R	point in 3D	4R	point in 4D (homogeneous)
2I	integer pair	IorP	integer or pointer
{ }	list		
[ ]	default value (other than the "default default")		
( )	arithmetic expression		
*	repeat the following (a list follows)		

Figure E.1 Key to type abbreviations used in prescriptions.

For further information, the appropriate version of the IGES specification should be consulted.

Entity	Form	Type	Prescription
0			null
100			circular arc
		R	z displacement
		2R	centre
		2R	start
		2R	terminate point
102			composite curve
		N	number of constituents
		N*{P}	constituent
104			conic arc
	0		determined by parameters
	1		elliptic
	2		hyperbolic
	3		parabolic
		6*{R}	coefficient
		R	z displacement
		2R	start
		2R	terminate point
106			copious data
	1		data pairs in a plane
	11		2D linear path
		I	interpretation flag
		N	number of tuples
		R	z displacement
		N*{2R}	data point in 2D
106			copious data
	2		data triples in a plane
	12		3D linear path
		I	interpretation flag
		N	number of tuples
		N*{3R}	data point in 3D
106			copious data
	3		data sextuples in a plane
	13		piecewise linear string
		I	interpretation flag
		N	number of tuples
		N*{3R	data point
		3R	vector
		}	
106			copious data
	20		centreline through points
	21		centreline through circle centres
		I	interpretation flag
		N	number of tuples
		R	z displacement
		N*{2R}	data point in 2d
106			copious data
	31		section: solid
	32		section: solid pairs
	33		section: solid + dash
	34		section: solid quads
	35		section: solid + dash + solid
	36		section: dash
	37		section: solid + perpendicular solid
	38		section: solid + perpendicular dash
		I	interpretation flag
		N	number of tuples
		R	z displacement
		N*{2R}	data point in 2D
106			copious data
	40		witness line
		I	interpretation flag
		N	number of tuples
		R	z displacement
		N*{2R}	data point in 2D
106			copious data
	63		simple closed planar curve
		I	interpretation flag
		N	number of tuples
		R	z displacement
		N*{2R}	data point in 2D
108			plane
	1		bounded void
			unbounded
	−1		bounded void
		4*{R}	coefficient
		P	closed curve
		3R	location point for display symbol
		R	size for display symbol
110			line
		3R	start point
		3R	terminate point
112			parametric spline curve
		I	spline type
		I	continuity
		I	dimension
		N	number of segments
		(N+1)*{R}	break point
		(3*N)*{4R}	polynomial coefficient
		4R	X derivative
		4R	Y derivative
		4R	Z derivative
114			parametric spline surface
		I	spline boundary type
		I	patch type
		NU	number of U segments
		NV	number of V segments
		(NU+1)*{R}	U break point
		(NV+1)*{R}	V break point

Entity	Form	Prescription	
		NU*{	
		(NV*48)*{R}	coefficient
		48*{R}	derivative
		}	
		((NV+1)*48)*{R}	derivative
116		point	
		3R	location
		P	display symbol
118		ruled surface	
	0	equal relative arc length	
	1	equal relative parametric values	
		2*{P}	curve
		I	direction flag
		I	developable surface flag
120		surface of revolution	
		P	axis of revolution
		P	generatrix
		R	start angle
		R	terminate angle
122		tabulated cylinder	
		P	directrix
		3R	terminate point of generatrix
123		direction	
		R	ratio wrt X
		R	ratio wrt Y
		R	ratio wrt Z [0.0]
124		transformation matrix	
	0	orthonormal rotate/translate	
	1	orthonormal rotate–reflect/translate	
	10	cartesian co-ordinates	
	11	cylindrical co-ordinates	
	12	spherical co-ordinates	
		3*{4R}	row
125		flash	
	0	defined by referenced entity	
	1	circular	
	2	rectangle	
	3	doughnut	
	4	canoe	
		2R	reference point
		2R	size parameter
		R	rotation
		P	referenced entity
126		rational B-spline curve	
	0	determined by parameters	
	1	line	
	2	circular arc	
	3	elliptical arc	
	4	parabolic arc	
	5	hyperbolic arc	

Entity	Form	Prescription	
		NK	upper index of sum
		NM	degree of basis functions
		I	planar flag
		I	closed flag
		I	periodic flag
		(NK+NM+2)*{R}	knot
		(NK+1)*{R}	weight
		(NK+1)*{3R}	control point
		2*{R}	parameter value
		3R	unit normal
128		rational B-spline surface	
	0	determined by parameters	
	1	plane	
	2	right circular cylinder	
	3	cone	
	4	sphere	
	5	torus	
	6	surface of revolution	
	7	tabulated cylinder	
	8	ruled surface	
	9	general quadric surface	
		NK1	upper index of first sum
		NK2	upper index of second sum
		NM1	degree of first set of basis functions
		NM2	degree of second set of basis functions
		I	closed in first direction
		I	closed in second direction
		I	rational flag
		I	periodic in first direction
		I	periodic in second direction
		(NK1+NM1+2)*{R}	first direction knot
		(NK2+NM2+2)*{R}	second direction knot
		((NK1+1)*(NK2+1))*{R}	weight
		((NK1+1)*(NK2+1))*{3R}	control point
		2*{R}	first direction parameter
		2*{R}	second direction parameter
130		offset curve	
		P	host curve
		I	distance flag
		P	what function
		I	dimension
		I	tapered offset type
		R	first offset distance
		R	first parameter value
		R	second offset distance
		R	second parameter value

Entity	Form	Prescription
		3R unit normal
		2*{R} parameter value
132		connect point
		3R location
		P display symbol
		I type flag
		I function flag
		S function identifier
		P pointer for function identifier
		S function name
		P pointer for function name
		I unique identifier
		I function code
		I swappable
		P owner subfigure
134		node
		3R node co-ordinate
		P nodal displacement co-ordinate system
136		finite element
		I topology type
		N number of nodes
		N*{P} node defining element
		S element type name
138		nodal displacement and rotation
		NA number of analyses
		NA*{P} analysis case description
		N number of nodes
		N*{ node
		I node number
		P node
		NA*{3R translation
		3R rotation
		}
		}
140		offset surface
		3R offset indicator
		R distance
		P host surface
141		boundary
		I representation type
		I preferred representation
		P untrimmed host surface
		N number of curves
		N*{P curve
		I orientation flag
		NK number of associated parameter space curves
		NK*{P} associated parameter space curve
		}
142		curve on a parametric surface
		I creation method
		P host surface
		P host entity
		P curve
		I preferred representation
143		bounded surface
		I representation type
		P untrimmed host surface
		N number of boundaries
		N*{P} boundary
144		trimmed parametric surface
		P host surface
		I outer boundary flag
		N number of inner boundaries
		P outer boundary
		N*{P} inner boundary
146		FEM nodal results
	0–34	(see Table D.1)
		P analysis case description
		I subcase
		R time
		NV number of values
		N number of nodes
		N*{ node
		I node number
		P node
		NV*{R} value
		}
148		FEM element results
	0–34	(see Table D.1)
		P description
		I subcase
		R time
		I number of values per location
		I reporting flag
		NE number of elements
		NE*{ element
		I element number
		P FEM element
		I topology type
		I number of layers
		I data layer flag
		N number of locations
		N*{I} location
		N number of values
		N*{R} value
		}
150		block
		R X length
		R Y length

Entity	Form	Prescription	
		R	Z length
		3R	corner point
		3R	local X axis [x_axis]
		3R	local Z axis [z_axis]
152		right angular wedge	
		R	X length at Y 0
		R	Y length
		R	Z length
		R	X length at Y max
		3R	corner point
		3R	local X axis [x_axis]
		3R	local Z axis [z_axis]
154		right circular cylinder	
		R	height
		R	radius
		3R	face centre
		3R	axis [z_axis]
156		right circular cone frustrum	
		R	height
		R	base radius
		R	top radius
		3R	base centre
		3R	axis [z_axis]
158		sphere	
		R	radius
		3R	centre
160		torus	
		R	major radius
		R	minor radius
		3R	centre
		3R	axis
162		solid of revolution	
	0	closed to axis	
	1	closed to itself	
		P	curve to be revolved
		R	fraction of full rotation [1.0]
		3R	centre
		3R	axis
164		solid of linear extrusion	
		P	closed curve
		R	extrusion length
		3R	extrusion direction
168		ellipsoid	
		R	X length
		R	Y length
		R	Z length
		3R	centre
		3R	local X axis [x_axis]
		3R	local Z axis [z_axis]
180		Boolean tree	
		N	post order notation length
		N*{IorP}	operator or operand
182		selected component	
		P	host solid
		3R	point in or on desired component
184		solid assembly	
		N	number of items
		N*{P}	item
		N*{P}	matrix
186		manifold solid B-rep object	
		P	shell
		L	orientation
		N	number of void shells
		N*{P	void shell
		L	orientation
		}	
190		plane surface	
	0	unparameterized	
		P	location
		P	normal
190		plane surface	
	1	parameterized	
		P	location
		P	normal
		P	reference direction
192		right circular cylindrical surface	
	0	unparameterized	
		P	location
		P	axis
		R	radius
192		right circular cylindrical surface	
	1	parameterized	
		P	location
		P	axis
		R	radius
		P	reference direction
194		right circular conical surface	
	0	unparameterized	
		P	location
		P	axis
		R	radius
		R	semi angle
194		right circular conical surface	
	1	parameterized	
		P	location
		P	axis
		R	radius
		R	semi angle
		P	reference direction
196		spherical surface	
	0	unparameterized	

Entity	Form	Type	Prescription
		P	location
		R	radius
196			spherical surface
	1		parameterized
		P	location
		R	radius
		P	axis
		P	reference direction
198			toroidal surface
	0		unparameterized
		P	location
		P	axis
		R	major radius
		R	minor radius
198			toroidal surface
	1		parameterized
		P	location
		P	axis
		R	major radius
		R	minor radius
		P	reference direction
202			angular dimension
		P	general note
		2*{P}	witness line
		2R	vertex
		R	radius
		2*{P}	leader
204			curve dimension
		P	general note
		2*{P}	curve
		2*{P}	leader
		2*{P}	witness line
206			diameter dimension
		P	general note
		2*{P}	leader
		2R	centre
208			flag note
		3R	lower left co-ordinate
		R	rotation angle
		P	general note
		N	number of leaders
		N*{P}	leader
210			general label
		P	general note
		N	number of leaders
		N*{P}	leader
212			general note
	0		simple note (default)
	1		dual stack
	2		imbedded font change
	3		superscript
	4		subscript
	5		super-/subscript
	6		multiple stack/left justified
	7		multiple stack/centre justified
	8		multiple stack/right justified
	100		simple fraction
	101		dual stack fraction
	102		imbedded font change/ double fraction
	105		super-/subscript fraction
		N	number of text strings
		N*{	text string
		I	number of characters
		R	box width
		R	box height
		IorP	font characteristic [1]
		R	slant angle [half_pi]
		R	rotation angle
		I	mirror
		I	rotate internal
		3R	start point
		S	text string
		}	
213			new general note
		R	text containment area width
		R	text containment area height
		I	justification code
		3R	text containment area position
		R	rotation angle
		3R	base line position
		I	normal interline spacing
		N	number of text strings
		N*{	text string
		I	fixed or variable
		R	character width
		R	character height
		R	intercharacter spacing
		R	interline spacing
		I	font style
		R	character angle
		S	control code
		I	number of characters
		R	box width
		R	box height
		IorP	character set [1]
		R	slant angle [half_pi]
		R	rotation angle
		I	mirror
		I	rotate internal
		3R	start point
		S	text string
		}	

Entity	Form	Prescription
214		leader arrow
	1	wedge
	2	triangle
	3	filled triangle
	4	no arrowhead
	5	circle
	6	filled circle
	7	rectangle
	8	filled rectangle
	9	slash
	10	integral sign
	11	open triangle
	12	dimension origin
		`N` number of segments
		`R` arrowhead height
		`R` arrowhead width
		`R` z depth
		`2R` arrowhead coordinates
		`N*{2R}` tail segment
216		linear dimension
	0	undetermined
	1	diameter form
	2	radius form
		`P` general note
		`2*{P}` leader
		`2*{P}` witness line
218		ordinate dimension
	0	witness line or leader
		`P` general note
		`P` witness line or leader
218		ordinate dimension
	1	witness line and leader
		`P` general note
		`P` witness line
		`P` leader
220		point dimension
		`P` general note
		`P` leader
		`P` enclosing curve
222		radius dimension
		0 single leader format
		`P` general note
		`P` leader
		`2R` centre
222		radius dimension
	1	multiple leader format
		`P` general note
		`P` first leader
		`2R` centre
		`P` second leader

Entity	Form	Prescription
228		general symbol
	0	general symbol
	1	datum feature
	2	datum target
	3	feature control frame
5001..9999		implementor defined
		`P` general note
		`N` number of geometry
		`N*{P}` geometry
		`N` number of leaders
		`N*{P}` leader
230		sectioned area
	0	standard crosshatching
	1	inverted crosshatching
		`P` boundary curve
		`I` line pattern
		`3R` mandatory point
		`R` separation
		`R` angle [quarter_pi]
		`N` number of island curves
		`N*{Pl}` island curve
302		associativity definition
		`NC` number of class definitions
		`NC*{I` back pointer flag
		`I` ordered
		`N` items per entry
		`N*{I}` data type
		}
304		line font definition
	1	subfigure pointer
		`I` orientation
		`P` display template
		`R` arc length distance
		`R` scale factor
304		line font definition
	2	repeating pattern
		`N` number of segments
		`N*{R}` segment length
		`S` blanked pattern
306		macro definition
		`S` MACRO
		`I` entity type
		`S` ENDM
308		subfigure definition
		`I` depth
		`S` name
		`N` number of entities
		`N*{P}` entity
310		text font definition
		`I` font number
		`S` font name

```
Entity Form   Prescription
              IorP        superseded font
              I           grid units
              N           number of characters
              N*{         character
                 I        ASCII code
                 2I       origin
                 NM       number of pen
                          moves
                 NM*{     pen move
                     I    pen up flag
                     2I   location
                    }
                }
312           text display template
        0     absolute co-ordinates
        1     incremental co-ordinates
              R     box height
              R     box width
              IorP  font characteristic [1]
              R     slant angle [half_pi]
              R     rotation angle
              I     mirror flag
              I     rotate internal flag
              3R    increment
314           colour definition
              3*{R} percentage of full intensity
              S     name
316           units data
              N       number of units
              N*{
                 S    type
                 S    unit
                 R    scale factor
                }
320           network subfigure definition
              I       depth
              S       name
              N       number of entities
              N*{P}   entity
              I       which type
              S       primary reference
                      designator name
              P       primary reference
                      designator
              N       number of children
              N*{P}   child
322           attribute table definition
        0     one-to-many
              S       table name
              I       list type
              NA      number of attributes
              NA*{    attribute
                  I   what type
```

```
Entity Form   Prescription
                  I value data type
                  I value count
                 }
322           attribute table definition
              1           one-to-one
              S           table name
              I           list type
              NA          number of attributes
              NA*{        attribute
                  I       what type
                  I       value data type
                  N       value count
                  N*{A}   value count
                 }
322           attribute table definition
        2     one-to-many with text display
              template
              S           table name
              I           list type
              NA          number of attributes
              NA*{        attribute
                  I       what type
                  I       value data type
                  N       value count
                  N*{A    value
                      P   text display template
                     }
                 }
402           associativity instance
        1     unordered group with backpointers
        7     unordered group without
              backpointers
       14     ordered group with backpointers
       15     ordered group without
              backpointers
              N       number of entries
              N*{P}   grouped entity
402           associativity instance
        2     external logical reference file index
       12     external reference file index
              N       number of index entries
              N*{S    external reference symbolic
                      name
                 P    internal entity
                }
402           associativity instance
        3     views visible
              NV        number of views visible
              N         number of entities
              NV*{P}    view
              N*{P}     entity
```

Entity	Form	Prescription	
402		associativity instance	
	4	views visible, colour, line weight	
		NV	number of views visible
		N	number of entities
		NV*{P	view
		I	line font
		P	line font definition
		IorP	colour
		I	line weight
		}	
		N*{P}	entity
402		associativity instance	
	5	entity label display	
		N	number of labels
		N*{P	view
		3R	text location
		P	leader
		I	level number
		P	labelled entity
		}	
402		associativity instance	
	6	view list	
		I	number of views
		N	number of visible entities
		P	view
		N*{P}	entity
402		associativity instance	
	8	signal string	
		NS	number of signal names
		NC	number of connect nodes
		NL	number of entities in logical signal
		NP	number of entities in physical signal
		NS*{S}	signal name
		NC*{P}	connect node
		NL*{P}	logical signal entity
		NP*{P}	physical signal entity
402		associativity instance	
	9	single parent associativity	
		I	number of parents
		N	number of children
		P	parent
		N*{P}	child
402		associativity instance	
	10	text node	
		N	number of points
		I	number of text descriptions
		N*{P}	point
		R	box height
		R	box width
		IorP	font characteristic [1]

Entity	Form	Prescription	
		R	slant angle [half_pi]
		R	rotation angle
		I	mirror flag
		I	rotate internal flag
402		associativity instance	
	11	connect node	
		NP	number of points
		N	number of data
		NP*{P}	point
		N*{A}	data
402		associativity instance	
	13	dimensioned geometry	
		I	number of dimensions
		N	number of geometry
		P	dimension
		N*{P}	geometry
402		associativity instance	
	16	planar group	
		I	number of matrices
		N	number of planar entities
		P	matrix
		N*{P}	planar entity
402		associativity instance	
	18	flow	
		I	number of context flags
		NF	number of flows
		NC	number of connect points
		NJ	number of joints
		NFN	number of flow names
		NT	number of text display templates
		NX	number of continuations
		I	flow type
		I	function flag
		NF*{P}	flow
		NC*{P}	connect point
		NJ*{P}	join
		NFN*{S}	flow name
		NT*{P}	text display template
		NX*{P}	continuation
402		associativity instance	
	19	segmented views visible	
		N	number of blocks
		N*{	
		P	view
		R	breakpoint parameter
		I	display flag
		IorP	colour
		IorP	line font
		I	line weight
		}	

Entity	Form		Prescription
402			associativity instance
	20		piping flow
		I	number of context flags
		NF	number of piping flows
		NC	number of connect points
		NJ	number of joins
		NFN	number of flow names
		NT	number of text display templates
		NX	number of continuations
		I	flow type
		I	function flag
		NF*{P}	piping flow
		NC*{P}	connect point
		NJ*{P}	join
		NFN*{S}	flow name
		NT*{P}	text display template
		NX*{P}	continuation
402			associativity instance
	21		dimensioned geometry
		ND	number of dimensions
		NG	number of geometry
		P	dimension
		I	dimension orientation
		R	angle
		NG*{	
		P	geometry
		I	dimension location
		3R	point on geometry
		}	
402			associativity instance
5001..9999			implementor defined (302 required)
404			drawing
	0		translate and scale
		N	number of views
		N*{P	view
		2R	origin
		}	
		N	number of annotation entities
		N*{P}	annotation entity
404			drawing
	1		translate, scale and rotate
		N	number of views
		N*{P	view
		2R	origin
		R	orientation angle
		}	
		N	number of annotation entities
		N*{P}	annotation entity

Entity	Form		Prescription
406			property
	1		definition levels
		N	number of property values
		N*{I}	level
406			property
	2		region restriction
		I	number of property values
		I	electrical vias restriction
		I	electrical components restriction
		I	electrical circuitry restriction
406			property
	3		level function
		I	number of property values
		I	function description code
		S	function description
406			property
	4		region fill property
		I	number of property values
		I	fill code
		P	obsolete section
406			property
	5		line widening
		I	number of property values
		R	metallization width
		I	cornering codes
		I	extension flag
		I	justification flag
		R	extension value
406			property
		6	drilled hole
		I	number of property values
		R	drill diameter
		R	finish diameter
		I	plating flag
		I	lower layer
		I	higher layer
406			property
	7		reference designator
		I	number of property values
		S	reference designator
406			property
	8		pin number
		I	number of property values
		S	pin number
406			property
	9		part number
		I	number of property values
		S	generic part number
		S	MIL STD part number
		S	vendor part number
		S	internal part number
406			property
	10		hierarchy

Entity	Form	Prescription	
		I	number of property values
		I	line font
		I	view
		I	entity level
		I	blank status
		I	line weight
		I	colour
406		property	
	11	tabular data	
		I	number of property values
		I	property type
		ND	number of dependent variables
		NI	number of independent variables
		ND*{I}	type of independent variable
		NI*{I}	number of values for independent variable
		NI*sigma*{R}	values of independent variables
		ND*sigma*{R}	values of dependent variables
406		property	
	12	external reference file list	
		N	number of property values
		N*{S}	file name
406		property	
	13	nominal size	
		I	number of property values
		R	nominal size value
		S	nominal size name
		S	relevant engineering standard
406		property	
	14	flow line specification	
		N	number of property values
		S	primary flow line specification
		N*{S}	modifier
406		property	
	15	name	
		I	number of property values
		S	name
406		property	
	16	drawing size	
		I	number of property values
		2R	drawing extent
406		property	
	17	drawing units	
		I	number of property values
		I	units flag
		S	units name
406		property	
	18	character spacing	
		I	number of property values
		R	spacing percentage
406		property	
	19	line font	
		I	number of property values [1]
		I	line font pattern
406		property	
	20	highlight	
		I	number of property values [1]
		I	highlight flag
406		property	
	21	pick	
		I	number of property values [1]
		I	pick flag
406		property	
	22	uniform rectangular grid	
		I	number of property values [9]
		I	finite flag
		I	line or point flag
		I	weighted flag
		2R	grid point
		R	X spacing
		R	Y spacing
		I	number in X
		I	number in Y
406		property	
	23	associativity group type	
		I	number of property values
		I	associativity type
		S	unique name
406		property	
	24	level to PWB layer map	
		I	number of property values
		ND	number of level to layer definitions
		ND*{	
		I	exchange file level number
		S	sending system native level
		I	physical layer number
		S	exchange file level identification
		}	
406		property	
	25	PWB artwork stackup	
		I	number of property values
		S	artwork stackup identification
		N	number of level values
		N*{I}	PWB level
406		property	
	26	drilled hole	
		I	number of property values [3]
		R	drill diameter

Entity	Form	Prescription	
		R	finish diameter
		I	function code
406		property	
	27	generic data	
		I	number of property values
		S	name
		N	number of type and value pairs
		N*{	
		I	data type
		A	value
		}	
406		property	
	28	dimension units	
		I	number of property values
		I	relative position
		I	units indicator
		I	character set
		S	formatting string
		I	fraction flag
		I	precision or denominator
406		property	
	29	dimension tolerance	
		I	number of property values
		I	secondary tolerance flag
		I	tolerance flag
		I	tolerance placement [2]
		R	upper tolerance
		R	lower tolerance
		L	suppress sign
		I	fraction flag
		I	display precision
406		property	
	30	dimension display data	
		I	number of property values
		I	dimension type
		I	preferred label position
		I	character set
		S	text
		I	decimal point symbol
		R	witness line angle [half_pi]
		I	text alignment
		I	text level
		I	preferred text placement
		I	arrowhead orientation
		R	primary dimension initial value
		N	number of supplemental notes
		N*{	
		I	relative position
		I	start index
		I	end index
		}	

Entity	Form	Prescription	
406		property	
	31	basic dimension	
		I	number of property values
		2R	lower left
		2R	lower right
		2R	upper right
		2R	upper left
406		property	
5001..9999		implementor defined	
	N	number of property values	
		N*{A}	value
408		singular subfigure instance	
		P	subfigure definition
		3R	translation
		R	scale factor [1.0]
410		view	
	0	orthographic parallel	
		I	view number
		R	scale factor [1.0]
		6*{P}	view volume
410		view	
	1	perspective	
		I	view number
		R	scale factor
		3R	view plane normal
		3R	view reference point
		3R	centre of projection
		3R	view up vector
		R	view plane distance
		R	left clipping window co-ordinate
		R	right clipping window co-ordinate
		R	bottom clipping window co-ordinate
		R	top clipping window co-ordinate
		I	depth clipping indicator
		R	back clipping window co-ordinate
		R	front clipping window co-ordinate
412		rectangular array subfigure instance	
		P	definition
		R	scale factor [1.0]
		3R	lower left
		I	number of columns
		I	number of rows
		R	horizontal separation
		R	vertical separation
		R	rotation
		N	number of do/donts
		I	do/dont flag

Entity	Form	Prescription
		N*{I} position
414		circular array subfigure instance
		P definition
		I maximum number of locations
		3R centre
		R radius
		R start angle
		R delta angle
		N number of do/donts
		I do/dont flag
		N*{I} position
416		external reference
	0	one definition from library file
	2	external logical references
		S file identifier
		S symbolic name
416		external reference
	1	full referenced library file
		S file identifier
416		external reference
	3	native system subfigure
		S symbolic name
416		external reference
	4	native library subfigure
		S library name
		S symbolic name
418		nodal load and constraint
		N number of cases
		I load or constraint flag
		P node
		N*{P} load case definition
420		network subfigure instance
		P definition
		3R translation
		3R scale [1.0, xs, xs]
		I which type
		S primary reference designator name
		P primary reference designator
		N number of children
		N*{P} child
422		attribute table instance
	0	single row
422		attribute table instance
	1	multiple row

Entity	Form	Prescription
430		solid instance
		P solid
502		vertex
		N number of vertices
		N*{3R} co-ordinate
504		edge
		N number of edge tuples
		N*{
		P curve
		P start vertex
		I start vertex index
		P terminate vertex
		I terminate vertex index
		}
508		loop
		N number of edge tuples
		N*{
		I edge type
		P vertex or edge
		I list index
		L orientation
		NK number of parameter space curve
		NK*{
		L isoparametric flag
		P parameter space curve
		}
		}
510		face
		P underlying surface
		N number of loops
		L outer loop flag
		N*{P} loop
514		shell
		N number of faces
		N*{P face
		L orientation
		}
600..999		implementor defined macro instance
5001..9999		implementor defined macro instance
10000..99999		implementor defined macro instance

Appendix F

VDA-FS entity prescriptions

This appendix presents the attribute names and types for VDA-FS entities, using the same notation as that used for IGES, given in Figure E.1. The VDA-FS specifications should be consulted for further information in interpreting the attributes.

Entity	Prescription
POINT	point 3R location
PSET	point sequence N number of points N*{3R} location
MDI	point vector sequence N number of pairs N*{3R data point 3R vector }

CURVE	parametric spline curve
	N number of segments
	N+1)*{R} parameter value
	N*{
	NO polynomial order
	NO*{R} polynomial X coefficient
	NO*{R} polynomial Y coefficient
	NO*{R} polynomial Z coefficient
	}

SURF	parametric spline surface
	NS number of S patches
	NT number of T patches
	(NS+1)*{R} parameter in S value
	(NT+1)*{R} parameter in T value
	(NS*NT)*{
	NU polynomial order in U direction
	NV polynomial order in V direction
	(NU*NV)*{R} X coefficient
	(NU*NV)*{R} Y coefficient
	(NU*NV)*{R} Z coefficient
	}

HEADER	header identification I number of continuation lines
BEGINSET	start of set combination (no attributes)
ENDSET	end of set combination (no attributes)
END	end of file identification (no attributes)
CIRCLE	circle or circular arc 3R centre R radius 2*{3R} orthonormal vector R start angle R finish angle

Entity	Prescription	
CONS	curve on surface	
	`P`	surface
	`P`	curve
	`2*{R}`	curve parameter
	`N`	number of segments
	`(N+1)*{R}`	global parameter
	`N*{`	
	`NP`	polynomial order
	`NP*{R}`	s coefficient
	`NP*{R}`	t coefficient
	`}`	
FACE	trimmed surface	
	`P`	surface
	`N`	number of closed curve sequence
	`N*{`	
	`NN`	number of CONS entities
	`NN*{`	
	`P`	curve on surface
	`2R`	parameter
	`}`	
	`}`	
TOP	topology	
	`N`	number of pairs of touching patches
	`N*{`	
	`2*{`	
	`P`	surface
	`NN`	number of CONS entities
	`NN*{`	
	`P`	curve
	`2R`	parameter
	`}`	
	`}`	
	`I`	continuity
	`}`	
GROUP	group associativity	
	`N`	number of elements
	`N*{P}`	element name
TMAT	transformation matrix	
	`4*{3R}`	row
TLIST	transformation list	
	`P`	transformation matrix
	`N`	number of elements
	`N*{P}`	element name

Appendix G

SET sub-block prescriptions

This appendix presents the attribute names and types for SET entities, using the notation given in Figure G.1. Similarly, the SET specifications should be consulted for further information in interpreting the attributes.

I	integer	Nchar	integer used as a counter
R	real	P	pointer
coord	depends on dictionary value	C	character string

{ }	list
[ ]	optional
*	repeat the following (a list follows)

Figure G.1 Key to type abbreviations used in SET prescriptions.

Entity	Prescription
1	point co-ordinates
	coord co-ordinates
2	straight line segment
	coord start
	coord end
3	polygonal line
	I dimension
	I closed
	{coord} vertex
4	vector components
	coord components
10	parameters of a circular arc
	R radius
	R start angle
	R end angle
	coord centre

Entity	Prescription
11	parameters of an elliptical arc
	R semi major radius
	R semi minor radius
	R start angle
	R end angle
	R rotation angle
	coord centre
12	parameters of a parabolic arc
	R parabola coefficient
	R start parameter
	R end parameter
	R rotation angle
	coord vertex
13	parameters of a hyperbolic arc
	2R hyperbola coefficient

Entity	Prescription	
	R	start angle
	R	end angle
	R	rotation angle
	coord	centre
14	coefficients of a conic	
	6R	conic coefficients (a,b,c,d,e,f)
20	start of a parametric polynomial curve	
	I	dimension
	I	generation method
	I	continuity
	R	initial parameter value
	R	final parameter value
	[R]	offset distance
21	definition of a polynomial curve segment	
	R	initial parameter value
	R	final parameter value
	NI	number of coefficients in X
	NI*R	coefficients
	NI	number of coefficients in Y
	NI*R	coefficients
	[NI]	number of coefficients in Z
	[NI*R]	coefficients
	[NI]	number of coefficients for homogeneous co-ordinate
	[NI*R]	coefficients
30	cartesian definition of a plane	
	4R	plane coefficients (a,b,c,d)
31	parametric definition of a plane	
	coord	point on plane
	coord	first vector
	coord	second vector
40	start of a parametric polynomial surface	
	I	generation method
	I	continuity
	I	topology
	I	number of patches in U
	I	number of patches in V
	R	lower U parameter
	R	upper U parameter
	R	lower V parameter
	R	upper V parameter
	[R]	offset distance
41	definition of a polynomial patch	
	R	lower limit of U
	R	upper limit of U
	R	lower limit of V
	R	upper limit of V
	NC	number of columns of coefficients in X
	NR	number of rows of coefficients in X

Entity	Prescription	
	(NC*NR)*R	X coefficients
	NC	number of columns of coefficients in Y
	NR	number of rows of coefficients in Y
	(NC*NR)*R	Y coefficients
	NC	number of columns of coefficients in Z
	NR	number of rows of coefficients in Z
	(NC*NR)*R	Z coefficients
	[NC]	number of columns of homogeneous coefficients
	[NR]	number of rows of homogeneous coefficients
	[(NC*NR)*R]	homogeneous coefficients
120	definition of a composite curve	
	I	dimension
	{	
	P	curve
	I	direction
	}	
	[R]	offset distance
130	parametric definition of a surface of revolution	
	R	lower parameter
	R	upper parameter
	R	start angle
	R	end angle
	P	generation curve
	Pcoord	point on axis of revolution
	Pcoord	unit vector on axis of revolution
	Pcoord	reference point for angular parameter
131	parametric definition of a tabulated cylinder	
	R	lower parameter for generating curve
	R	upper parameter for generating curve
	R	lower parameter for directrix
	R	upper parameter for directrix
	P	generating curve
	Pcoord	directrix unit vector
132	parametric definition of a ruled surface	
	R	lower parameter for first curve
	R	upper parameter for first curve

Entity	Prescription	
	R	lower parameter for second curve
	R	upper parameter for second curve
	R	lower parameter for directrix
	R	upper parameter for directrix
	P	first curve
	P	second curve
	[IP]	corresponding points method
	[I]	direction of curves (end to end)
216	projection method	
	I	projection type
	R	scale
	[2R	conic projection vertex or
	[R]]	projection direction vector
220	definition of drawing space	
	R	width
	R	height
	I	unit
	R	scale factor
	I	angle unit
	[C]	format name
221	display of view on a drawing	
	IP	view
	2R	lower left corner
	2R	size of viewing window
	R	rotation angle
	I	border display
	I	text display
	I	axes display
	[2R]	virtual window lower left corner (pan and zoom)
	[2R]	virtual window size
251	graphical representation of a list of blocks	
	{P}	wire frame entity
252	graphical representation of a block in a list of views	
	I	display flag
	{IP}	view
253	linear plot (pen)	
	I	unit length
	R	offset
	{	
	I	pen up
	R	abscissa
	}	
254	linear positioning of symbols	
	I	plot unit
	R	offset
	{IP	symbol
	R	height
	R	width
	R	abscissa
	}	
255	association of ASCII code and character definition	
	I	decimal value of ASCII code
	I	total points in X
	I	total points in Y
	I	filling code
	{I	pen up
	2IR	co-ordinates
	}	
257	graphical representation of a symbol	
	I	total points in X
	I	total points in Y
	I	filling code
	2I	reference co-ordinates
	{I	pen up
	2IR	co-ordinates
	}	
258	graphical representation of a wireframe entity (variable line font)	
	IP	display flag
	R	curve width
	I	curve justification
	R	first change parameter value
	{IP	line font
	R	parameter value
	}	
301	reference system's transformation matrix	
	4R	rotation matrix coefficients
	[5R]	rotation matrix coefficients
	[2R [R]]	translation vector
310	windowing	
	I	clipping type
	4R	spherical clipping values
	or	
	2R	depth clipping values
	or	
	5R	cylindrical clipping values
	or	
	6R	box clipping values
312	list of levels	
	{I}	level
313	definition of a 3D transformation	
	3R	scale
	[3I	symmetry

Entity	Prescription
	[9R rotation matrix coefficients
	[3R translation vector
	]]]
314	general matrix of a 3D transformation
	9R rotation matrix coefficients
	[3R] translation vector
315	definition of a 2D transformation
	2R scale
	[2I symmetry
	[4R rotation matrix coefficients
	[2R translation vector
	]]]
316	general matrix of 2D transformation
	4R rotation matrix of coefficients
	[2R] translation vector
317	translation
	2R [R] vector components
400	TNODE parameters
	IC name
401	text description
	2R text start point
	C first line of text
	{C} subsequent lines of text
402	dimension text
	I type of text
	2R text start point
	C text
406	linear dimension line
	I arrow symbol placement
	2R first point
	2R second point
	{2R} subsequent points
407	angular dimension line
	I arrow symbol placement
	2R arc centre
	R radius
	R start angle
	R end angle
	[R extension angle of arc
	{2R} broken lines extension
	]
408	witness line
	2R first segment start
	2R first segment end
	{2R 2R} subsequent segments start and end
409	box (framework)
	I symbol type
	R length or diameter of symbol
	R width of symbol
	2R symbol origin
	R rotation angle
410	axis lines
	2R first segment start
	2R first segment end
	{2R 2R} subsequent segments start and end
430	cross-hatching
	I material code
	2R first segment start
	2R first segment end
	{2R 2R} subsequent segments start and end
500	call definition
	I whether duplicated
	PC object to be called
510	associativity defined by SET
	I nature of associativity
	I whether shared
	I individualization
	I block type
	P first block
	{P} subsequent blocks
515	associativity towards representation
	I nature of associativity
	{P} entity
519	instance name and technical reference
	C name
	[C] technical specification
520	attributes defined by SET
	C name
	I attribute type
	{IRC} values
521	attributes defined by user
	C name
	I attribute type
	{IRC} values
602	logical link between two CNODES
	P first Cnode
	P second Cnode
610	list of component terminals
	{C} Cnode
80xx	user-defined
	(none specified by SET)
9900	file identification
	C SET version
	C interface software version
	C CADCAM software name
	C CADCAM software version
	C company name

Entity	Prescription	
	C	division name
	C	department name
	C	designer name
9901	date	
	I	year
	I	month
	I	day
	[I	hour
	I	minute
	I	second
	]	
9902	comments	
	C	text
9905	model space definition	
	I	dimension
	R	scale
	I	unit code

Entity	Prescription	
	R	scale factor
	I	angle unit
	4R [2R]	window
9911	last sub-block in a series of sub-blocks	
	(no attributes)	
9920	initialization of dictionary parameters	
	[I	dictionary parameter number
	any	new default value
	]	
9930	header block descriptor	
	I	class number
	I	parameter number of the dictionary
	{I}	parameter number of the dictionary

Appendix H

Addresses and telephone numbers

AECMA
88 Boulevard Malesherbes
F-75008 Paris
France

Phone: + 33 1 4563 8285

Aerospatiale
Direction Technique
92152 Suresnes
France

Phone: +33 1 772 11 22

American Society of Mechanical Engineers
United Engineering Center
345 East 47th Street
New York
NY 10017
USA

Association GOSET
107–111 Avenue Georges Clemenceau
F-92000
France

CAD-CAM Data Exchange Technical Centre
University of Leeds
Woodhouse Lane
Leeds LS2 9JT
UK

CTN,
HQ AFLC LMSC/SBC
Wright-Patterson Air Force Base
OH 45433
USA

Defense Single Stock Point
Commanding Officer
Naval Publications and Forms Center
5801 Tabor Avenue
Philadelphia
PA 19120
USA

Department of Mechanical Engineering
University of Leeds
Woodhouse Lane
Leeds LS2 9JT
UK

Phone: + 44 113 233 2113

Department of Trade and Industry
London
UK

Phone: + 44 171 215 1532

EDIF Users' Group
Design Automation Department
Texas Instruments
PO Box 225474
MS3668
Dallas
TX 75265
USA

EDiSTONE Books
West Winds
Inkford Brook
Wythall
Birmingham B47 6DA
UK

Phone: + 44 1564 823707

Electronic Industries Association
2001 Pennsylvania Avenue N.W.
Washington DC 20006
USA

European Committee for IT&T Testing and Certification
Rue de Stassart 36
B-1050 Brussels

Phone: + 32 2 519 68 11

IGES/PDES Organization
National Institute of Standards and Technology
Gaithersburg
MD 20899
USA

Information Analysis and Dissemination Center
WL/MTI
Wright-Patterson Air Force Base
OH 45433-6533
USA

Information Geometers Ltd
47 Stockers Avenue
Winchester SO22 5LB
UK

Phone: + 44 1962 841996

IPC
1717 Howard Street
Evanston
IL 60202
USA

Mantech Technology Transfer Center
WL/MTX
Wright-Patterson Air Force Base
OH 45433-6533
USA

Materials Laboratory (AFWAL/MLTC)
Air Force Wright Aeronautical Laboratories
Wright-Patterson Air Force Base
OH 45433
USA

National Bureau of Standards,
Gaithersburg
MD 20899
USA

National Economic Development Office
Millbank Tower
Millbank
London SW1P 4QX
UK

National Institute of Standards and Technology
Gaithersburg
MD 20899
USA

National Technical Information Service
US Department of Commerce
Springfield
VA 22161
USA

NCGA IPO Administrator
National Computer Graphics Association
2722 Merrilee Drive
Suite 200
Fairfax
VA 22031
USA

Standardized Documents Order Desk
Building 4D
700 Robbins Avenue
Philadelphia
PA 19111-5094
USA

Verband der Automobilindustrie e. V,
D-6000 Frankfurt am Main
Westendstraße 61
Germany

Textual notes

1. The authors of IGES 1.0 were Walt Braithwaite, Phil Kennicott and Roger Nagel. As a result of this achievement, they received the Jacquard Memorial Award of the Association for Integrated Manufacturing Technology. This is the highest award of the association, and had been awarded only thirteen times in its 25-year history.
2. Clearly, not all of the initiatives have been included in this book: two projects in the USA, GMAP and RAMP, are notable absentees.
3. Anyone may submit a processor to a laboratory for conformance testing, but needs to be able to provide certain information about the processor; consequently, although a vendor is the most likely client, others (such as user groups) are also possible. The term "client" is used in the remainder of this text to encompass all possibilities.
4. Some sections in an International Standard are prescribed by ISO. A standard is divided into clauses which may, in turn, be partitioned into subclauses. Annexes may also be included. By its very nature, the numbered clauses of a standard are normative; that is, they define requirements to be fulfilled. An annex may be designated as normative or informative; the latter means that it contains no requirements, but provides background material to help the reader to understand the normative clauses and annexes.
5. NCGA is a non-profit organization of individuals and major companies in the computer graphics industry. The association supports the development and use of graphics standards by participating in ANSI and ISO and administering the Graphics Performance Characterization (GPC) Committee. This committee is developing standardized benchmarks for measuring performance of graphics subsystems.
6. In this chapter, only a few sites and initiatives have been mentioned. Although the authors believe that these are representative, clearly not all sites undertaking work in this area can be or have been included. If you know of other work, the authors would be interested to learn of details, either by electronic mail or conventional post at the addresses given in the foreword. Details of work in countries not mentioned in Section 11.10

would be particularly welcome, as this could be used as a point of contact between that country and those with centres established already.

7. The Japanese migration strategy mentioned earlier uses several EXPRESS schemata: one for IGES, those that are part of STEP itself, one to represent the information structures in the CAD system, and one to effect the conversions. Such EXPRESS can then be processed by software to produce large parts of the translator automatically.

References

Useful addresses and telephone numbers for obtaining these documents are given in Appendix H; see also Appendix B.

[1] AECMA 1984. *AECMA recommendations for geometry data exchange*. Paris: AECMA.

[2] Altemueller, J. 1988. *The STEP file structure*. ISO TC184/SC4/WG1 document N279, September (STEP Tokyo draft proposal, annex B). Gaithersburg, Maryland: NIST.

[3] Altemueller, J. 1988. *Mapping from EXPRESS to physical file structure*. ISO TC184/SC4/WG1 document N280, September (STEP Tokyo draft proposal, annex C). Gaithersburg, Maryland: NIST.

[4] Altmueller, J. 1988. *Development support*. ISO TC184/SC4/WG1 document N289, October (STEP Tokyo draft proposal, annex E). Gaithersburg, Maryland: NIST.

[5] Ames, A. L. & S. K. Fletcher 1985. *Making IGES work for CAD/CAM data exchange*. Paper presented at Autofact'85. Report MS85-1110. Dearborn, Michigan: Society of Manufacturing Engineers.

[6] ANSI 1973 x3.39 and x3.22. New York: ANSI.

[7] ANSI 1977. *Printed board description in digital form*. (ANSI/IPC-D-350B). Evanston, Illinois: IPC.

[8] Arnold, D. B. & D. A. Duce 1990. *ISO standards for computer graphics: the first generation*. London: Butterworth.

[9] Ashworth, M. J. 1993. *A STEP-based information strategy and tools for engineering organizations*. PhD thesis, Department of Mechanical Engineering, University of Leeds.

[10] Baker, G., J. C. Kelly, W. Conroy 1992. *IGES 5.1 recommended practices guide*. Gaithersburg, Maryland: IGES/PDES Organization.

[11] Bézier, P. 1986. *The mathematical basis of the UNISURF CAD system*. London: Butterworth.

[12] Blacker, B. K. (ed.)1988. *CADCAM data exchange: equipment vendors' response to supporting IGES, VDA-FS and SET*. London: Society of Motor Manufacturers and Traders.

[13] Blacker, K. (ed.) 1989. Advancing the application of IGES. London: Society of Motor Manufacturers and Traders.

[14] Blacker, K. 1994. *The basics of electronic data exchange*. Birmingham: EDiSTONE Books.

[15] Bloor, M. S., S. A. Corne, J. D. Cash 1990. Conformance testing for CADCAM interfaces. In [31], XXXI:1–25.

[16] Bloor, M. S., J. R. Dodsworth, J. Owen 1984. *CAD interchange of data: guidelines for the use of IGES. Phase 1: two-dimensional scale and non-scale drawings with text*. London: National Economic Development Office.

[17] Bloor, M. S., J. R. Dodsworth, J. Owen, P. F. Stewart 1986. *CAD interchange of data: guidelines for the use of IGES. Phase 2: two-dimensional drawings with associated data*. Leeds: CADDETC.

[18] Bono, P. R. 1986. Graphics standards: guest editor's introduction. *IEEE Computer Graphics and Applications*, **6**(8), 12–16.

[19] Booz, Allen and Hamilton. *Final report on the PDDI project*. AFWAL/MLPL, Wright Patterson Air Force Base, OH 45433, USA.

[20] Brinksma, E. 1986. A tutorial on LOTOS. See: [40] 171–94.

[21] Brown, H. 1989. Standards for structured documents. *Computer Journal*, **32**(6), 505–14.

[22] Brown, J. M. 1992. *The design and properties of surfaces generated using partial differential equations*. PhD thesis, Departments of Applied Mathematical Studies and Mechanical Engineering, The University of Leeds.

[23] BS 5750-1 1987. *British Standard 5750 Quality systems Part 1: Specification for design/development, production, installation and servicing*. London: British Standards Institution.

[24] CAM-I 1980. *An interface between geometric modellers and application programs*. CAM-I Report R-80-GM-04.

[25] Carson, G. S. 1993. Introduction to the Computer Graphics Reference Model. *Computer Graphics* **27**(2), 108–19.

[26] CGI 1991. ISO/IEC 9636. Information technology – Computer graphics – Interfacing techniques for dialogues with graphical devices (CGI) – functional specification. Part 1: Overview, profiles and conformance. Part 2: Control. Part 3: Output. Part 4: Segments. Part 5: Input and echoing. Part 6 : Raster. Geneva: ISO.

[27] CGM 1992. ISO/IEC 8632. Information technology – Computer graphics – Metafile for the storage and transfer of picture description information (CGM). Part 1: Functional specification. Part 2: Character encoding. Part 3: Binary encoding. Part 4: Clear text encoding. Geneva: ISO.

[28] CGRM 1992. ISO/IEC 11072. Information technology – Computer graphics – Computer Graphics Reference Model. Geneva: ISO.

[29] Clarke, C. G., M. S. Bloor, S. A. Roberts, P. E. Preece, M. P. Atkinson, P. Bailey 1989. *Engineering database evaluation project: report for the Computer Facilities Technical Advisory Group (CFTAG) of the Science and Engineering Research Council*. SERC project number DB22082. Available from Department of Mechanical Engineering, The University of Leeds.

[30] Coons, S. A. 1967. *Surfaces for computer-aided design of space forms*. Project MAC Report TR-MAC-41. Boston: Massachusetts Institute of Technology.

[31] CTS2 1990. *European conference on conformance testing and certification in information technology and telecommunications*. Proceedings of the conference held at Hotel Palace, Brussels, Belgium, from 13 to 15 June 1990, sponsored by the CEC and EFTA Secretariat.

[32] CTS2 1991. CTS2 Technical Report 3: Abstract test cases: rationale, definitions and examples. Document ISO TC184/SC4/WG6/N32, 15 April. Gaithersburg, Maryland: NIST.

[33] CTS2 1992. CTS2 Technical Report 4: Final report on global test concepts. Document ISO TC184/SC4/WG6/N47, 21 January. Gaithersburg, Maryland: NIST.

[34] CTS2 1993. CTS2 Technical Report 5: SET schematics (service procedures, software and test models) and Generic model for a Test Method Control Board. Document ISO TC184/SC4/WG6/N80, 15 January. Gaithersburg, Maryland: NIST.

[35] CTS2 1993. CTS-II-CAD/CAM project: final report and exploitation report. Document ISO TC184/SC4/WG6/N81, 1 November. Gaithersburg, Maryland: NIST.

[36] Dahl, O.-J., E. W. Dijkstra, C. A. R. Hoare 1972. *Structured programming*. London: Academic Press.

[37] Dannenberg, L.& H. Nowacki 1985. Approximate conversion of surface representations with polynomial bases. *Computer-aided Geometric Design* **2**(1–3), 123–31.

[38] Davies, M. S., J. P. Fowler, H. J. Kahn, R. Lau 1990. STEP/EDIF compatibility study. Report AT/90/0001. Leeds: CADDETC.

[39] Day, A. 1988. Information models. ISO TC184/SC4/WG1 document N288, 25 November. (STEP Tokyo draft proposal, annex D). Gaithersburg, Maryland: NIST.

[40] Diaz, M. (ed.) 1986. Protocol specification, testing, and verification, vol. V. *Proceedings of the IFIP WG 6.1 fifth international workshop on protocol specification, testing, and verification*. Amsterdam: North-Holland.

[41] DTI 1987. *Through MAP to CIM*. London: Department of Trade and Industry.

[42] DTI 1990. *Open systems in manufacturing: product data exchange*. London: Department of Trade and Industry.

[43] DTI 1991. *Open systems in manufacturing: an introduction to the manufacturing message specification*. London: Department of Trade and Industry.

[44] DTI 1992. *CALS: the executive guide*. London: Department of Trade and Industry.

[45] Duce, D. A. & F.釋ernó 1993. A formal specification of a graphics system in the framework of the computer graphics reference model. *Computer Graphics Forum* **12**(1), 3–20.

[46] ECITC 1993. *The ECITC guide to IT&T testing and certification*. Brussels: ECITC.

[47] EDIF 1985. *Electronic Design Interchange Format, version 1 1 1*. Dallas, Texas: EDIF Users' Group.

[48] EDIF 1993. *Electronic Design Interchange Format, version 3 0 0*. Washington, DC: Electronic Industries Association.

[49] EN 29001 1987. *EN 29001. Quality systems – model for quality assurance in design/development, production, installation and servicing*. Brussels: CEN.

[50] Encarnação, J., R. Schuster, E. Vöge (eds) 1986. *Product data interfaces in CAD/CAM applications*. Berlin and New York: Springer.
[51] Ewald, R. H. & R. Fryer (eds) 1978. Final report of the GSPC state of the art subcommittee. *Computer Graphics* **12**(1–2), 14–169.
[52] Faux, I. D. & M. J. Pratt 1983. *Computational geometry for design and manufacture*. Chichester, England: Ellis Horwood.
[53] Fleming, J. 1988. Test methods – conformance. ISO TC184/SC4/WG1 document N285, 31 October (STEP Tokyo draft proposal, clause 5). Geneva: ISO.
[54] Fuhr, R. D. 1985. Conversion between rational B-spline and parametric spline representations. B-7137-DF-WP-632/ZZ3/1 (Boeing Commercial Airplane Company, Seattle, WA, USA). Available from: National Institute of Standards and Technology, Gaithersburg, MD 20899, USA.
[55] Gielingh, W. F. 1993. European R&D related to product data technology. *Proceedings of the PDT Days 1993 conference*, 63–76. VDI Berichte 1096. Verein Deutscher Ingenieure.
[56] GKS 1985. ISO 7942. Information processing systems – Computer graphics – Graphical Kernel System (GKS) functional description. Geneva: ISO.
[57] GKS language bindings 1988. ISO 8651. Information processing systems – Computer graphics – Graphical Kernel System (GKS) language bindings. Part 1: Fortran. Part 2: Pascal. Part 3: Ada. Part 4: C. Geneva: ISO.
[58] GKS-3D 1988. ISO 8805. Information processing systems – Computer graphics – Graphical Kernel System for Three Dimensions (GKS-3D) functional description. Geneva: ISO.
[59] GKS-3D language bindings 1991. ISO 8806. Information technology – Computer graphics – Graphical Kernel System for Three Dimensions (GKS-3D) language bindings. Part 1: Fortran. Part 4: C. Geneva: ISO.
[60] Goult, R. J. & P. A. Sherar (eds) 1990. *Improving the performance of neutral file data transfers. Esprit Project 322 (CAD*I)*, vol. 6. Heidelberg: Springer.
[61] Grabowski, H. (ed.) 1991. *Advanced modelling for CAD/CAM systems. Esprit Project 322 (CAD*I)*, vol. 7. Heidelberg: Springer.
[62] Grabowski, H. & R. Glatz 1987. IGES model comparison system: a tool for testing and validating IGES processors. *IEEE Computer Graphics and Applications* **7**(11), 47–57.
[63] GSPC 1977. Status report of the Graphics Standards Planning Committee of ACM SigGraph. *Computer Graphics* **11**(3).
[64] GSPC 1979. Status report of the Graphics Standards Planning Committee of ACM SigGraph. *Computer Graphics* **13**(3).
[65] Halvorson, H. & M. E. Palmer 1995. *STEP application protocols: status and summary report*. Gaithersburg, Maryland: IGES/PDES Organization.
[66] Henderson, L. R. 1991. CALS as a solution for digital delivery of technical documents. *Computer-Aided Design* **23**(4), 252–6.
[67] Henderson, L. R. & A. M. Mumford 1990. *The Computer Graphics Metafile*. London: Butterworth.
[68] Higa, K., M. Morrison, J. Morrison, O. R. L. Sheng 1992. An object-oriented methodology for knowledge base/database coupling. *Communications of the ACM* **35**(6), 99–113.
[69] Hopgood, F. R. A., D. A. Duce, J. R. Gallop, D. C. Sutcliffe 1986. *Introduction to the Graphical Kernel System (GKS)*. London: Academic Press.

[70] Huau, P. 1991. *User guide for the SET standard.* Document GDT-F/91-002. Association GOSET.

[71] IDEF0 1981. *ICAM Architecture Part II, Vol. IV – Function Modeling Manual (IDEF0).* Report number AFWAL-TR-81-4023, vol. IV. June. Mantech Technology Transfer Center, WL/MTX.

[72] IDEF1 1981. *ICAM Architecture Part II, Vol. V – Information Modeling Manual (IDEF1).* Report number AFWAL-TR-81-4023, vol. V. June. Information Analysis and Dissemination Center, WL/MTI.

[73] IDEF1X 1985. *Integrated Information Support System (IISS). Vol. V – Common data model subsystem. Part 4 – Information Modeling Manual – IDEF1 Extended.* Report number AFWAL-TR-86-4006, vol. V, part 4. November. Mantech Technology Transfer Center, WL/MTX.

[74] IDEF2 1981. *ICAM Architecture Part II, Vol. VI – Dynamics Modeling Manual (IDEF2).* Report number AFWAL-TR-81-4023, vol. VI. June. Information Analysis and Dissemination Center, WL/MTI.

[75] IGES 1980. *The Initial Graphics Exchange Specification (IGES) version 1.0.* NBSIR 80-1978 (R). 1 January. Gaithersburg, Maryland: National Bureau of Standards.

[76] IGES 1981. *ANSI Y14.26M: Digital representation for communication of product definition data.* New York: American Society of Mechanical Engineers.

[77] IGES 1983. *Initial Graphics Exchange Specification (IGES) version 2.0.* Springfield, Virginia: National Technical Information Service.

[78] IGES 1986. *Initial Graphics Exchange Specification (IGES) version 3.0.* Gaithersburg, Maryland: National Bureau of Standards.

[79] IGES 1988. *The Initial Graphics Exchange Specification (IGES) version 4.0.* NISTIR 88-3813. June. Springfield, Virginia: National Technical Information Service.

[80] IGES 1990. *The Initial Graphics Exchange Specification (IGES) version 5.0.* NISTIR 4412. Gaithersburg, Maryland: National Institute of Standards and Technology.

[81] IGES 1991. *The Initial Graphics Exchange Specification (IGES) version 5.1.* Gaithersburg, Maryland: IGES/PDES Organization.

[82] ISO 1987. *ISO Technical Report 9007: Information processing systems – Concepts and terminology for the conceptual schema and the information base.* Geneva: ISO.

[83] ISO 9001 1987. *ISO standard 9001: Quality systems – Model for quality assurance in design/development, production, installation and servicing.* Geneva: ISO.

[84] ISO 1991. ISO/IEC 9646. Information technology – Open Systems Interconnection – Conformance testing methodology and framework. Part 1: General concepts. Part 2: Abstract test suite specification. Part 3: The Tree and Tabular Combined Notation (TTCN). Part 4: Test realization. Part 5: Requirements on test laboratories and clients for the conformance assessment process. Geneva: ISO.

[85] ISO 1992. ISO/IEC Technical Report 10000. Information technology – Framework and taxonomy of International Standardized Profiles. Part 1: Framework. Part 2: Taxonomy of OSI Profiles. Geneva: ISO.

[86] Kroszynski, U. I., B. Palstroem, E. Trostmann, E. G. Schlechtendahl 1989.

Geometric data transfer between CAD systems: solid models. *IEEE Computer Graphics and Applications* **9**(5), 57–71.

[87] Lammers, M. S. 1991. *CALS Test Network handbook*. Available from: CTN, HQ AFLC LMSC/SBC, Wright-Patterson AFB, Ohio 45433, USA

[88] Lamport, L. 1986. *LATEX: a document preparation system*. Reading, Mass.: Addison-Wesley.

[89] Liewald, M. H. 1985. Initial graphics exchange specification: successes and evolution. *Computers and Graphics* **9**(1), 47–50.

[90] Liewald, M. H. & P. R. Kennicott 1982. Intersystem data transfer via IGES. *IEEE Computer Graphics and Applications* **2**(3,) 55–63.

[91] Marca, D. A. & C. L. McGowan 1988. *SADT: Structured Analysis and Design Technique*. New York: McGraw-Hill.

[92] Mayer, R. J., T. Cullinane, P. S. deWitte, W. B. Knappenberger, B. Perakath, M. S. Wells 1992. *IDEF3 process description capture method report*. Report number AL-TR-1992-0057. May. Mantech Technology Transfer Center, WL/MTX.

[93] Mayer, R.J., A. Keen, M. S. Wells 1992. *IDEF4 object-oriented design method manual*. Report number AL-TR-1992-0056. May. Mantech Technology Transfer Center, WL/MTX.

[94] McKay, A. & J. Owen 1995. *STEP: a framework for product modelling*. Winchester, UK: Information Geometers.

[95] McKay, A., M. S. Bloor, J. Owen 1994. Application protocols: a position paper. Presented at in the MICAD Conference and proceedings, Paris, March 1994 and published as *International journal of CADCAM and computer graphics*, **9** (3), 377–89.

[96] MIL-D-28000 1987. *Digital representation for communication of product data: IGES application subsets*. Defense single stock point, Commanding officer, Naval Publications and Forms Center, 5801 Tabor Avenue, Philadelphia, PA 19120, USA.

[97] MIL-D-28000A 1992. *Digital representation for communication of product data: IGES application subsets and IGES application protocols*. Philadelphia, Pennsylvania: Standardized Documents Order Desk.

[98] MIL-STD-1840 () *Automated interchange of technical information*. Philadelphia, Pennsylvania: Standardized Documents Order Desk.

[99] Mumford, A. M. 1987. Why care about the Graphics Graphics Metafile? *Computer-Aided Design* **19**(8), 425–30.

[100] Mumford, A. M. 1991. Taking a positive approach to standards (guest editorial). *Computer-Aided Design* **23**(4), 235–6.

[101] Naur, P. (ed.) et al. 1960. Report on the algorithmic language Algol 60. *Numersiche Mathematik* **2**, 106–36.

[102] ODA 1989. ISO/IEC 8613. Information processing – Text and office systems – Office Document Architecture (ODA) and interchange format. Part 1: Introduction and general principles. Part 2: Document Structures. Part 4: Document profile. Part 6: Character content architectures. Part 7: Raster graphics content architectures. Part 8: Geometric. Part 10: Formal specifications. Geneva: ISO.

[103] ODIF 1989. ISO/IEC 8613. Information processing – Text and office systems – Office Document Architecture (ODA) and interchange format. Part 5: Office Document Interchange Format (ODIF). Geneva: ISO.

[104] Olle, T.W, H. G. Sol, A. A. Verrijn-Stuart (eds) 1982. *Information systems design methodologies: a comparative review*. Amsterdam: North-Holland.

[105] OSI 1989. ISO/IEC 7498. Information processing systems – Open Systems Interconnection – Basic Reference Model. Part 1: Basic reference model and Addendum 1: Connectionless-mode transmission. Part 2: Security architecture. Part 3: Naming and addressing. Part 4: Management framework. Geneva: ISO.

[106] Owen, J. 1988. Introduction, scope and definitions. ISO TC184/SC4/WG1 document N283, 31 October (STEP Tokyo draft proposal, clauses 0–3). Gaithersburg, Maryland: NIST.

[107] Owen, J. 1990. *Standards for product data exchange and conformance testing*. Paper presented at the joint EuroGraphics and British Computer Society Displays Group state of the art seminar "Systems integration and data exchange", London, 28 February.

[108] Owen, J. 1993. *STEP: an introduction*.Winchester, UK: Information Geometers.

[109] Owen, J. 1993. STEP: a critical technology for business. *Proceedings of the PDT Days 1993 conference*, 49–62. VDI Berichte 1096: Verein Deutscher Ingenieure.

[110] Owen, J. & M. S. Bloor 1987. Neutral formats for product data exchange: the current situation. *Computer-Aided Design* **19**(8), 436–43.

[111] Pahl, G. & W. Beitz 1984. *Engineering design*. London: Design Council.

[112] Palmer, M. E. 1993. *Guidelines for the development and approval of STEP application protocols, version 1.1*. ISO TC184/SC4/PMAG N103. 30 November. Gaithersburg, Maryland: NIST.

[113] PDDI 1984. *Product Definition Data Interface*. Report number SS 5601 30200. July. Materials Laboratory (AFWAL/MLTC).

[114] PHIGS 1989. *ISO/IEC 9592. Information processing systems – Computer graphics – Programmer's Hierarchical Interactive Graphics System (PHIGS)*. Part 1: Functional description. Part 2: Archive file format. Part 3: Clear-text encoding of archive file. Geneva: ISO.

[115] PHIGS language bindings 1990. *ISO 9593. Information processing systems – Computer graphics – Programmer's Hierarchical Interactive Graphics System (PHIGS) language bindings*. Part 1: Fortran. Part 3: Ada. Part 4: C. Geneva: ISO.

[116] PHIGS-PLUS 1992. *ISO/IEC 9592. Information processing systems – Computer graphics – Programmer's Hierarchical Interactive Graphics System (PHIGS)*. Part 4: Plus lumière und surfaces, PHIGS-PLUS. Geneva: ISO.

[117] Pierra, G. 1990. An object oriented approach to ensure portability of CAD standard parts libraries. *Proceedings of EuroGraphics 90*, 205–14. September 1990.

[118] Pitt, D. H. & D. Freestone 1990. The derivation of conformance tests from LOTOS specifications. *IEEE Transactions on Software Engineering* **16**(12), 1337–43.

[119] Pratt, M. J. 1985. IGES – the present state and future trends. *Computer-Aided Engineering Journal* **2**(4), 130–3.

[120] Pratt, M. J., R. J. Goult, M. J. Chinnery 1987. *Comparison and translation of surface representations with CADCAM systems*. London: Society of Motor Manufacturers and Traders.

[121] Raflik, M. & B. Pätzold (eds) 1990. *An approach to an engineering database (version 4.0)*. Esprit Project 322 (CAD*I) vol. 5. Heidelberg: Springer.
[122] Rehwald, P. 1985. VDA-FS – an interface to transfer surface description data between CAD systems. *Computers and Graphics* **9**(1), 69–70.
[123] Requicha, A. A. G. 1980. Representations for rigid solids; theory, methods and systems. *ACM Computing Surveys* **12**(4), 437–64.
[124] Requicha, A. A. G. & J. R. Rossignac 1992. Solid modeling and beyond. *IEEE Computer Graphics and Applications* **12**(5), 31–44.
[125] Requicha, A. A. G. & H. B. Voelker 1982. Solid modeling: a historical summary and contemporary assessment. *IEEE Computer Graphics and Applications* **2**(2), 9–24.
[126] Requicha, A.A.G. and Voelker, H.B. 1983. Solid modeling: current status and research directions. *IEEE Computer Graphics and Applications* **3**(7), 25–37.
[127] Ross, D. T. 1977. Structured analysis (SA): a language for communicating ideas. *IEEE Transactions on Software Engineering* **SE-3**(1), 16–34.
[128] Ross, D. T. & K. E. Schoman Jnr 1977. Structured analysis for requirements definition. *IEEE Transactions on Software Engineering* **SE-3**(1), 6–15.
[129] Rupp, C. G. et al. 1993. Integrated process capture and process analysis tools in support of business re-engineering applications. *CE & CALS Washington'92 conference proceedings*, 275–330.
[130] Schenk, D. 1988. Information modelling language EXPRESS. ISO TC184/SC4/WG1 document N287, 31 October (STEP Tokyo draft proposal, annex A). Gaithersburg, Maryland: NIST.
[131] Schenck, D. A. & P. R Wilson 1994. *Information modelling the Express way*. Oxford: Oxford University Press.
[132] Schlechtendahl, E. G. (ed.) 1988. *Specification of a CAD*I neutral file for CAD geometry, wireframes, surfaces, solids (version 3.3)*. Esprit Project 322 (CAD*I) vol. 1. Heidelberg: Springer.
[133] Schlechtendahl, E. G. (ed.) 1989. *CAD data transfer of solid models*. Esprit Project 322 (CAD*I) vol. 3. Heidelberg: Springer.
[134] Schuster, R. & R. Anderl 1986. CAD-Interfaces: a tool for integrating production processes. *Proceedings of MICAD'86*, 722–41.
[135] Schuster, R., D. Trippner, M. Endres (eds) 1990. *CAD*I drafting model*. Esprit Project 322 (CAD*I) vol. 4. Heidelberg: Springer.
[136] SET 1984. *Standard d'echange et de transfert, version 1.1*. Suresnes: Aerospatiale.
[137] SET: z68-300 1985. *Industrial automation – external representation of product definition data – data exchange and transfer standard specification version 85-08*. Paris: AFNOR.
[138] SET: z68-300 1989. *Industrial automation – external representation of product definition data – data exchange and transfer standard specification version 89-06*. Paris: AFNOR.
[139] SGML 1986. ISO 8879. *Information processing – Text and office systems – Standard Generalized Markup Language (SGML)*. Geneva: ISO.
[140] Shepherd, B. J. 1987. Graphics standards: issues and approaches (guest editorial). *Computer-aided Design* **19**,(8), 394–5.
[141] Smith, B. 1998. Classification methods. ISO TC184/SC4/WG1 document N286, 31 October (STEP Tokyo draft proposal, clause 6). Gaithersburg, Maryland: NIST.

[142] Smith, B. M. 1983. IGES: a key to CAD/CAM systems integration. *IEEE Computer Graphics and Applications* **3**(8), 78–83.

[143] Smith, B. M. et al. 1986. *A reporting of the PDES initiation activities*. Gaithersburg, Maryland: National Bureau of Standards.

[144] SPDL 1991. ISO/IEC Draft International Standard 10180. Information technology – Text communication – Standard Page Description Language (SPDL). Geneva: ISO.

[145] Sproull, R. F. 1982. Using program transformation to derive line-drawing algorithms. *ACM Transactions on Graphics* **1**(4), 259–73.

[146] STEP 1992. *Industrial automation systems and integration – Product data representation and exchange – Part 31: Conformance testing methodology and framework: General concepts*. ISO DIS 10303-31. ISO TC184/SC4 N157. Geneva: ISO.

[147] Sutcliffe, D. C., J. R. Gallop, R. Maybury, R. A. Day, D. A. Duce, G. J. Reynolds 1993. The ARGOSI classification scheme for graphics and networking applications. *Computer Graphics Forum* **12**(2), 123–30.

[148] Thomas, D., J. van Maanen, M. Mead (eds) 1989. *Specification for exchange of product analysis data*. Esprit Project 322 (CAD*I) vol. 2. Heidelberg: Springer.

[149] Trippner, D. H. Nowacki 1986. *Functional subsets of entities for STEP logical layer*. ISO TC184/SC4/WG1 document N60, March. Gaithersburg, Maryland: NIST.

[150] Tsichritzis, D. & A. Klug (eds) 1978. The ANSI/X3/SPARC DBMS framework: report of the study group on database management systems. *Information Systems* **3**, 173–91.

[151] VDA-FS 1984. *VDA surface interface version 1.0*. Frankfurt am Main: Verband der Automobilindustrie e.V.

[152] VDA-FS 1986. *DIN 66301*. Berlin: Deutsches Institut für Normung e. v.

[153] VDA-FS 1987. *VDA surface interface version 2.0*. Frankfurt am Main: Verband der Automobilindustrie e.V.

[154] VDA-IS 1987. *VDA IGES subset version 1.0*. Frankfurt am Main: Verband der Automobilindustrie e.V.

[155] VDA-IS 1989. *VDA IGES subset version 2.0*. Frankfurt am Main: Verband der Automobilindustrie e.V.

[156] VDA-PS 1987. *VDA Programmschnittstelle*. Frankfurt am Main: Verband der Automobilindustrie e.V.

[157] Verheijen, G. M. A. & J. van Bekkum 1982. NIAM: an information analysis method. See [104] (1982), 537–89.

[158] VHDL 1992. IEEE Draft Standard: language reference manual. IEEE P1076 1992A.

[159] Wang, W. & G. G. Grinstein. 1993. A survey of 3D solid reconstruction from 2D projection line drawings. *Computer Graphics Forum* **12**(2), 137–58.

[160] Weiss, J. 1986. *STEP functional requirements*. ISO TC184/SC4 N30 (2 May 1988). Gaithersburg, Maryland: NIST.

[161] Williams, A. L. 1988. *A guide to reading an IDEF1X model*. ISO TC184/SC4/WG1 document N273. Gaithersburg, Maryland: NIST.

[162] Williams, A. L. 1989. *Comparative analysis of the modeling languages IDEF1X, Express and NIAM*. ISO TC184/SC4/WG1 document N369. Gaithersburg, Maryland: NIST.

[163] Willis, R., & J. G. Nell 1992. *IGES/PDES Organization reference manual.* Fairfax, Virginia: NCGA IPO Administrator, National Computer Graphics Association.

[164] Wilson, P. R. 1987. Conic representations for shape description. *IEEE Computer Graphics and Applications* **7**(4), 23–30.

[165] Wilson, P. R. 1987. A short history of CAD data transfer standards. *IEEE Computer Graphics and Applications* **7**(6), 64–7.

[166] Wilson, P. R. 1987. Information and/or data? *IEEE Computer Graphics and Applications* **7**(11), 58–61.

[167] Wilson, P. R. 1987. Information modelling. *IEEE Computer Graphics and Applications* **7**(12), 65–7.

[168] Wilson, P. R. 1989. PDES STEPs forward. *IEEE Computer Graphics and Applications* **9**(2), 79–80.

[169] Wilson, P. R. 1990. STEP ballot results. *IEEE Computer Graphics and Applications* **10**(3), 79–82.

[170] Wilson, P. R. 1991. Standards: past tense and future perfect? *IEEE Computer Graphics and Applications* **11**(1), 44–8.

[171] Wilson, P. R. 1993. *Processing tools for EXPRESS.* Gaithersburg, Maryland: IGES/PDES Organization.

[172] Wilson, P. R. & P. R. Kennicott 1988. *ISO STEP baseline requirements document (IPIM).* ISO TC184/SC4/WG1 document N284, 27 October. (STEP Tokyo draft proposal, clause 4). Gaithersburg, Maryland: NIST.

[173] Wilson, P. R., I. D. Faux, M. C. Ostrowski, K. G. Pasquill 1985. Interfaces for data transfer between solid modeling systems. *IEEE Computer Graphics and Applications* **5**(1), 41–51; **5**(3), 83.

[174] Wirth, N. 1977. What can we do about the unnecessary diversity of notation for syntactic definitions? *Communications of the ACM* **20**(11), 822–3.

Index

abstract test case 143
  IGES example, extract 144
  STEP example, extract 187
accreditation 117, 150
  mutual recognition 143
  national requirements 141
activity modelling 212
AECMA 97
ANSI/SPARC 93, 170
ANSI Y14.26M 38
application protocols 131, 170
  see also STEP
ARGOSI 81
Autofact 40, 106

Booz, Allen and Hamilton 97
British royal family tree, extract 172
business-related standards 80

CADEX 162
CAD*I 160
CALS 161
CAM-I 27
certification 116
CGI 75
CGM 75
CGRM 201
chronology of development in product data viii
committees
  IPO committee structure (1983) 39
  IPO committee structure (1993) 39
  IPO Testing Project committee structure 112
  STEP committee structure 158
conformance testing 117, 133, 195
  IGES conformance testing activity diagrams 147, 148, 150
  STEP conformance testing activity diagram 189
context dependence 130
Core 73
CTS2 133, 143, 147

data 6
data exchange 15
  alleviated by use of a common system 16
  of graphical data 16
  using direct translators 16
  using neutral format 19
  using vendor formats 18
design process 10
DIN 66301 52
document-related standards 79

EDIF 76, 164
  syntax, extract 77
EDIFACT 80
Electrical and electronics-related standards 75, 164
ESPRIT-III STEP-related projects 163

European projects 163
examples
  Autofact 1981 demonstration part 40
  executable test case, extract 144
  EXPRESS schema 175
  EXPRESS-G 178
  EXPRESS-I 187
  IDEF0 147, 148, 150, 189
  IGES abstract test case, extract 144
  IGES file 31
  SET file 58
  STEP abstract test case, extract 187
  STEP physical file 180
  VDA-FS file 53
  WSN 214
exchange agreements 103
executable test case 135
  example, extract 144
EXPRESS 174
  example schema 175
  mapping to physical file 179
  software 196
EXPRESS-G 176
  example 178
  key to symbols 177
EXPRESS-I 178
  example 187

falsification testing 116
formalism 116, 201

GHOST 72
GINO 72
GKS 73
GKS Metafile 75
GKS-3D 74
GOSIP 79
graphics-related standards 71
  relationship with product data standards 71

ICAM Programme 211
IDEF0 212
  examples 147, 148, 150, 189
IDEF1X 155, 213
IGES 28, 63
  ANSI Y14.26M 38
  coverage 215
  directory entry fields 32
  entity prescriptions 223
  entity structure 29
  example file 31
  file format 30
  global section fields 30
  history of development 27, 49
  parameter data for circular arc 33
  recommended practices 98
  request for change 101
  subsets, original proposal (RFC 320A) 122
  version 1.0 28
  version 1.0 entities 35
  version 2.0 37
  version 3.0 63
  version 4.0 65
  version 5.0 67
  version 5.1 67
industrial and international co-ordination 166
information 6
information modelling 213
interoperability testing 117, 195
IPC 75
ISO 13584 196

language bindings 73, 193

magnetic tape
  IGES RP 6 100
  STEP requirements 181
MAP 79
matrix processing: IGES RPs 4 and 35 100
MBB part, used in the 1983 Autofact demonstration 41
membership of projects and consortia
  CAD I project 160
  Cadex project 162
  ECSI 78
  Pdes Inc 161
  PDTAG 166
MIL-D-SPECS 101, 123
MMS 79

NAVFAC tests 108
NEDO 95, 106
NIAM 155, 213

ODA 80
ODETTE 80
ODIF 80
Open Systems Interconnection 78
organisation protocol 202

performance testing 118
PHIGS 74
PDDI 59
 system architecture 61
 entities 60
PDES Inc 161
PDES initiation 159
problems
 communication between computer applications 3, 10
 demonstrations of errors in exchanges 45
 general data exchange 85, 89
 IGES 1.0 and 2.0 42
 neutral format transfer 86, 91
 STEP 198
product data 4, 7
product data model 4
product model 4
product data standards
 relationship with graphics standards 71
 centres of expertise 165

quality systems 143

robustness testing 118

SADT: see IDEF0
SDAI 193
SET 54, 69
 block classes 54
 block types and sub-block usage 56
 dictionary contents 55
 example file 58
 sub-block classes 54
 sub-block prescriptions 239
 sub-blocks 57
SGML 80
SMMT project: Advancing the Application of IGES 112
software
 automated development 196
 for conformance testing 138
 for testing 101, 109
software architectures
 Caddetc 139
 CADEX 163
 STEP (suggested) 197
SPDL 80
standard parts 195
STEP 151, 169
 abstract test case example 187
 application protocols, proposed 192
 application protocols, scope 186
 classes of parts 170
  abstract test suites 182
  application interpreted constructs 185
  application protocols 182
  conformance testing methodology and framework 188
  description methods 174
  implementation methods 178, 193
  integrated resources181
 conformance testing infrastructure 142
 EXPRESS *see* EXPRESS
 framework for product description and support 181, 200
 geometry-related schemata 183, 184
 history 153
 industrial adoption 201
 physical file
  example 180
  header section contents 179
 results of first ballot 156, 170
 status 173
 structure 169
 subsets, German proposal 121
 Tokyo Draft Proposal 156

topology schema 184
standardisation process 22
subsets
GKS levels 73
IGES, original proposal (RFC 320A) 122
MIL-D-28000A technical publications 124
MIL-D-28000A engineering drawings 125
MIL-D-28000A electrical and electronic 126
MIL-D-28000A NC 127
OSI 78, 131
STEP: German proposal 121
VDA-IS 128

testing methods:
IGES conformance testing 145
KfK 115
pre-processor conformance testing (physical file) 134
post-processor conformance testing (physical file) 135
Q-projects: method for data transfer 114
STEP conformance testing 189
testing programmes
CADDETC informal testing scope 108
IGES Committee and SAE 112
NAVFAC test drawing 110
NEDO 106
SMMT Q-projects scope 113
VDA-IS 113
TOP 79

user acceptance testing 117

validation 116
VDA-FS 51, 68
developers of version 1.0 52
developers of version 2.0 68
entity prescriptions 237
example file 53
version 1.0 51
version 2.0 68
VDA-IS 101, 128
VDA-PS 196
verification 116
VHDL 78

Wirth Syntax Notation (WSN) 213
use of 52, 78, 178

XBF 59

z68-300 58, 69